To Paul my good [illegible]
Best Wishes.
Dan McDermott
Oct 28, 2010.

EYE *of the* EXPLORER

Views of the Northern Pacific Railroad Survey, 1853–54

PAUL D. MCDERMOTT, RONALD E. GRIM, AND PHILIP MOBLEY

2010
Mountain Press Publishing Company
Missoula, Montana

Cover: "Northern Padific Railroad Survey Routes, 1853," map © 2010 by Philip Mobley, based on G. K. Warren map 1958, Library of Congress, Geography and Map Division; "Saint Paul" and "Mount Rainier viewed from near Steilacom," from Isaac I. Stevens, *Narrative and Final Report of Explorations for a Route for a Pacific Railroad*, vol. 12, bk.1 (Washington, DC: GPO, 1855)

Library of Congress Cataloging-in-Publication Data

McDermott, Paul D.
Eye of the explorer : views of the Northern Pacific Railroad survey, 1853–54 / Paul D. McDermott, Ronald E. Grim, and Philip Mobley.
p. cm.
Includes bibliographical references and index.
ISBN 978-0-87842-560-0 (cloth : alk. paper)
1. Pacific railroads—Explorations and surveys—History—19th century. 2. Northern Pacific Railroad Company—History. 3. Stevens, Isaac Ingalls, 1818–1862. 4. Explorers—West (U.S.)—History—19th century. 5. Surveyors—West (U.S.)—History—19th century. 6. West (U.S.)—Discovery and exploration. 7. West (U.S.)—Description and travel. 8. West (U.S.)—History—1848–1860. 9. Pacific railroads—Explorations and surveys—History—19th century—Pictorial works. 10. West (U.S.)—History—1848–1860—Pictorial works. I. Grim, Ronald E. II. Mobley, Philip, 1955– III. Title.
F593.M4745 2010
917.804'2—dc22
2010005033

PRINTED IN HONG KONG BY MANTEC PRODUCTION COMPANY

MP Mountain Press
PUBLISHING COMPANY
P.O. Box 2399 • Missoula, MT 59806 • 406-728-1900
800-234-5308 • info@mtnpress.com
www.mountain-press.com

This book is dedicated to the descendants of

Gustavus Sohon, especially Julian and Nancy Sohon.

And to our families, especially:

Gayl, David, and Shawn McDermott and beloved grandchildren Brijit, Liam, and Kairi

Michael Rockel and Alaina, Theron, and Myles Grim

Lynne, Erin, Becky, and Amanda Mobley

Contents

Acknowledgments

Many people have assisted us in creating this book. Several special people contributed photographic materials. We especially thank Kevin Hawkins for the fine photographs of Minnehaha Falls and Fort Snelling; Don Popejoy for the images and maps relating to Kettle Falls; and retired Capt. Dave Johanson for his wonderful shots of Ebey's Landing and the Madrona tree. Charles Baumgaertner and his wife, Ann, graciously spent two days photographing the Grand Coulee. His image was done in high dynamic range format. We are also grateful to Janell Best Rudel, who took time from a busy schedule to travel several hours to take photographs of Butte de Morale and the Sheyenne River. In addition, Dr. William Davis at the National Institute of Standards and Technology in Gaithersburg, Maryland, graciously assisted us in re-creating the night sky of August 25, 1853, showing Comet Klinkerfus.

Other images were furnished by Yale University Art Gallery, the University of Oregon, the Library of Congress, the National Archives and Records Administration, the Smithsonian Institution, Washington State Historical Library, Spokane Public Library, Boston Public Library, and the National Park Service. We would like to express our appreciation to the various staff members of these institutions for their assistance during our many years of research. Special thanks are extended to Christine Murphy at the Norman B. Leventhal Map Center for assistance with many last minute details.

Friends throughout the West contributed to our data collection and our travels. We especially thank the following people: Dr. Roland Grant and his wife, Kay, in Billings, Montana; Dr. Hazel Pflueger of Florence, Montana; David Nicandri, Director of the Washington State Historical Society in Tacoma, Washington; Bill Weikel, Missoula, Montana; and Ken Robeson, Great Falls, Montana. In memoriam, we thank Chuck Mead of Alberton, Montana, a gracious and thoughtful man who loved western history and his Montana homeland.

We also greatly appreciate the support provided by the editorial staff at Mountain Press in taking on this project and assisting us in bringing the book to fruition. Our editor, Gwen McKenna, was especially helpful in directing us to clarify our message and tighten our prose.

Finally, we want to acknowledge the Julian Sohon family, who generously shared family material regarding their ancestor, artist Gustavus Sohon. With much gratitude, we thank Julian and his wife, Nancy, for their support in developing this book.

PREFACE

In the early 1850s, the U.S. Congress authorized four surveys of the western United States to help determine the most practical route for a transcontinental railroad. Commonly known as the Pacific Railroad Surveys, these four traverses were conducted from 1853 to 1854. The northernmost survey, led by Isaac Stevens, traversed some of the same country first explored by Lewis and Clark at the beginning of the nineteenth century. Although none of the four routes was selected in its entirety for the first transcontinental railroad route, the surveys accumulated a vast amount of geographical data that resulted in the compilation of both a large, detailed composite map of the western United States and twelve volumes of invaluable reports describing the geography of the frontier West. The reports were generously illustrated with drawings of new botanical and zoological species observed in the West, as well as landscape views of the developed and wild country along the way.

In this book we reproduce, with commentary, the seventy illustrations included in Isaac Stevens's final report, which was published in 1860. Each of the original illustrations appears in color, accompanied by an examination of the image's historical and geographical importance. Quotations from the report, especially those specifically mentioning the creation of particular images, are scattered throughout the text. We also present newly compiled maps showing the location of the view in the context of modern regional geography, as well as modern photographs capturing views of the same landscapes today.

Part 1 is an introductory essay that provides historical, geographical, and artistic context for these illustrations. It addresses the Pacific Railroad Surveys and their significance for the exploration, mapping, and settlement of the Pacific Northwest; the major personalities associated with the survey (Isaac Ingalls Stevens, John Mullan, John Mix Stanley, and Gustavus Sohon); and commentary on the creation of these views and their relationship to the artistic representation of the American landscape, particularly in the West. Part 2 offers an overview of the survey route and explains the maps and the presentation of the illustrations. Part 3 is the Gallery of Views, the main body of this book. Here we examine the images one by one and discuss the details of each portion of the survey party's journey.

****

The impetus for this book began many years ago. As a child, Paul McDermott took frequent trips with his family across the West from Seattle to Chicago. As they crossed Montana on old US 10, he saw many wooden signs marking the Mullan Road, sparking his curiosity. He eventually learned that it was a 624-mile military wagon road built between Fort Benton and Fort Walla Walla, led by Lt. John Mullan, from 1859 to 1862. A lifelong fascination was born.

Dr. Ronald Grim, after finishing graduate work at the University of Maryland's Geography Department, where he and Paul met, rediscovered many documents and maps relating to Mullan's road-building project at the National Archives in Washington, D.C. This discovery led to a decades-long collaboration between Paul and Ron, as they dedicated themselves to learning more about Mullan and his work.

As they continued their pursuit, the pair became increasingly interested in the artist who accompanied Mullan on his expeditions, Gustavus Sohon. In the early 1990s, they searched for more information about him. With the aid of advancing computer technology, Paul and Ron were able to locate some of Sohon's descendants. After sending letters to six different persons, Paul received a phone call from Mr. Julian Sohon, who identified himself as a direct descendant of the artist. A close friendship soon developed, during which Julian and his family shared considerable family materials relating to Gustavus. Some of these materials have been incorporated into this book.

When this book project moved into its final stages, Philip Mobley joined the team as cartographer. Phil, an adjunct instructor at George Mason University and longtime friend and teaching colleague of Paul's, prepared all the modern maps that appear in this book. He and Paul traveled four thousand miles throughout the Pacific Northwest, exploring and photographing places important to Isaac Stevens's Northern Pacific Railroad Survey.

Part 1

Isaac Stevens's Northern Pacific Railroad Survey

The first half of the nineteenth century was a fascinating period in our nation's history. With the Louisiana Purchase in 1803 and acquisition of the Mexican and Oregon Territories during the 1840s, the United States greatly expanded its territory and resources. As settlers moved onto these lands, transportation routes were developed, first trails and navigable rivers, then roads, canals, and finally, railroads. Construction was preceded by extensive exploration and inventory appraisals, pioneered by the preliminary explorations of Lewis and Clark from 1803 to 1805.

Some fifty years after the Corps of Discovery's expedition, Isaac Stevens's Northern Pacific Railroad Survey provided intensive documentation of the northwestern United States, the same region Lewis and Clark had partially explored. While Lewis and Clark disproved the existence of a northwest river passage, the Stevens expedition proved the feasibility of another type of transportation artery—a railroad. These efforts were documented in word, map, and picture. For historians, this documentation provides substantial evidence of how the survey under Stevens's direction was accomplished, by whom, and with what results. Together, the textual and visual materials create a baseline by which natural and cultural change between the nineteenth and twentieth centuries can be evaluated.

The Surveys

The Pacific Railroad Surveys were conducted from 1853 to 1855—some fifty years after the expedition of Lewis and Clark—to determine the best route for the nation's first transcontinental railroad. The need for a railroad became evident as Americans began to migrate into the newly established California and Oregon Territories. A railroad was seen as the mechanism for integrating this land and its people with the East.

Because of the large capital expense involved, railroad proponents envisioned only a single rail line. Thus, there was considerable competition among parties advocating particular routes and terminal points for the railroad. In an attempt to find the best route and settle these arguments, Congress authorized and funded the Pacific Railroad Surveys. The surveys, whose approval was enacted March 2, 1853, were directed by Jefferson Davis, the Secretary of War under President Franklin Pierce.[1]

Five principal routes were selected for the surveys (see figure 1). Each was intended to be a traverse of a practically and politically feasible route. Four of these routes were east to west probes across the country, and the other was a south to north survey near the Pacific Coast. Each survey was undertaken by officers of the Army's Topographical Corps. These officers were directed to make their reports to the Bureau of Explorations and Surveys under the leadership of Maj. William Emory and later, Capt. Andrew Atkinson Humphreys. Their initial reports were to be completed by the end of 1853, as Davis was asked to submit his recommendations to Congress in January 1854. Even though the time window for this study was small, it was achievable if done solely as a reconnaissance.

In terms of the quantity and quality of geographic and scientific data collected, the most successful of the four east to west surveys was the northernmost, between the 47th and 49th parallels of latitude, headed by Maj. Isaac Stevens. Recently appointed as governor of the Washington Territory, Stevens was supported by Sen. Stephen Douglas of Illinois, a

Figure 1: Pacific Railroad Survey Routes, 1853 —Map by Philip Mobley

Figure 2: Northern Pacific Railroad Survey Routes, 1853
—Map by Philip Mobley

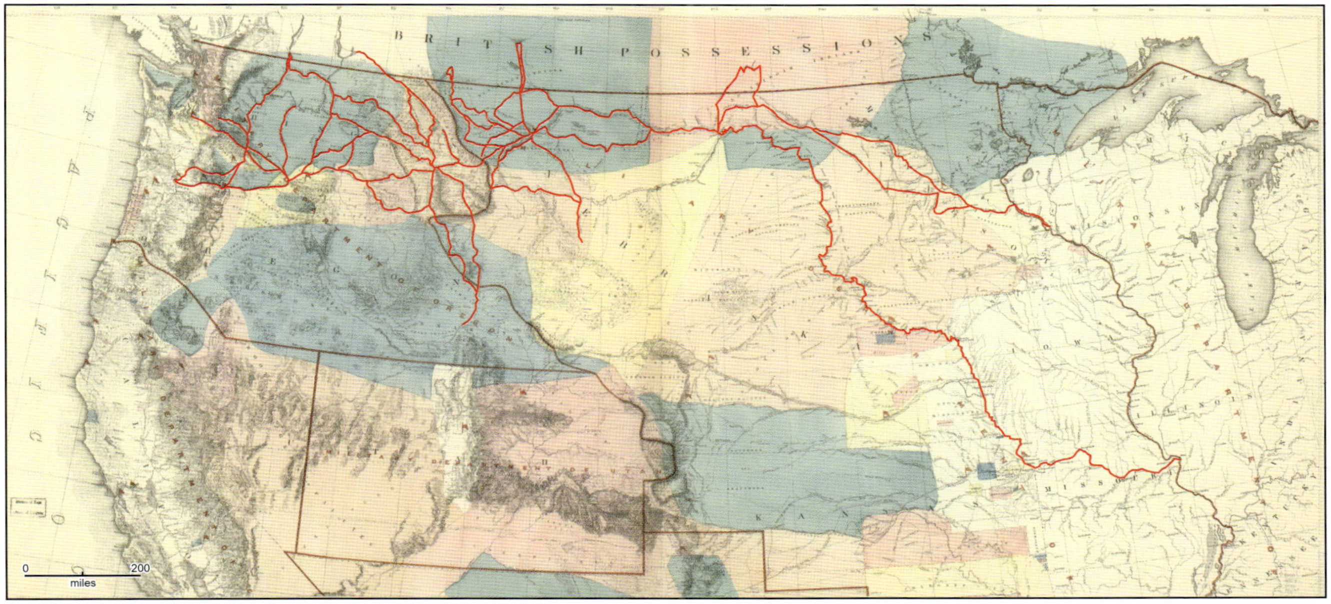

Source: G.K. Warren Map 1858, Library of Congress Geography and Map Division. Original scale 1:3,000,000

staunch advocate and promoter of the northern route. It was Stevens's task to explore a route from Duluth, Minnesota, to the Pacific Coast at Seattle (see figure 2).[2]

The scientific success of the survey was related to Stevens's penchant for organization and planning. Stevens believed that an expedition of this type must be preceded by extensive research. In this regard, he read a considerable amount of geographic and related literature prior to undertaking the project. This information was used in planning the project and selecting the most qualified personnel.

One of the major contributions of Stevens's survey was the variety of scientific data it gathered. Scientists were hired to study the route's geology, botany, topography, meteorology, and zoology. Supporting their efforts was a cadre of surveyors, soldiers, and artists. Ironically, the major personnel deficiency was the lack of experienced railroad engineers, and the survey failed to collect the precise engineering data required to build a railroad. This misdirection came from the top—major academic institutions and scientific societies pressured Secretary Davis to undertake comprehensive regional geographical investigations along the proposed routes at the expense of engineering data. In any case, the short deadline allowed insufficient time to collect such detailed information.

The actual survey work for the entire northern route, which was conducted under the overall direction of Isaac Stevens, was divided between two major groups working concurrently—one starting in the east, led by Stevens himself, and the other starting in the west, led by Capt. George B. McClellan, who later achieved notoriety as a Civil War general. The eastern group had the task of exploring and collecting data covering the bulk of the distance from the eastern terminal points at Duluth and St. Paul–Minneapolis west to the Cascade Range in Washington Territory. They joined the western party in the Puget Sound region. The goal of McClellan's group was to explore only the Cascade Mountains for potential passes. Unfortunately, McClellan, who had an aversion to snow, failed to complete his task or to obtain good data on snowfall. Interestingly, Davis used McClellan's faulty and incomplete data as a reason for rejecting the northern route. Eventually, however, parts of the northern route were used by three different railroads for portions of their lines to the Pacific—the Great Northern, the Northern Pacific, and the Chicago, Milwaukee & St. Paul Railroads.

Although the four east–west Pacific Railroad Surveys did not provide a conclusive solution to the best single route for a transcontinental railroad, they did identify several major corridors through which transcontinental railroads were later constructed. Perhaps a more important contribution, from an academic standpoint, was the accumulation of a corpus of geographical and scientific data on this largely unknown portion of the country. In fact, these surveys produced twelve volumes of published reports, augmented with numerous maps, scientific drawings, and landscape views. Robert Taft, in his classic study of western artists, *Artists and Illustrators of the Old West, 1850–1900,* affirmed that these reports "constitute probably the most important single contemporary source of knowledge on western geography and history, and their value is greatly enhanced by the inclusion of many beautiful plates in color of scenery, native inhabitants, fauna, and flora of the western country."[3]

While the maps published with these reports have received considerable attention by historians of exploration and cartography, it is the illustrations, especially the landscape drawings produced on Stevens's northern route of the Pacific Railroad Surveys, that are the subject of this book.[4] The landscape views, originally recorded as pencil and watercolor sketches, were printed as color lithographs when the final survey report was prepared for publication. These illustrations, which provided the first substantial visual image of the northwestern United States, were primarily the creations of two accomplished artists, John Mix Stanley and Gustavus Sohon, who accompanied Isaac Stevens and his survey party as they explored the area between the 47th and 49th parallels.

The Personalities

Stevens's survey party was blessed with many interesting personalities, some of whom also became important in the history of the Pacific Northwest. Besides Stevens himself, who provided the overall leadership for the expedition and was one of the major players in Washington Territory's politics during the 1850s, other notables included John Mullan, John Mix Stanley, and Gustavus Sohon.

Mullan, one of Stevens's most trusted officers on the expedition, was responsible for conducting numerous secondary explorations of the lands adjacent to the survey's main traverse. These experiences honed the foresight and leadership abilities he later applied to the construction of a strategic military wagon road between Fort Walla Walla, Washington Territory, and Fort Benton, Nebraska Territory, as well as the promotion of land sales and settlement of the region. Stanley, already a noted artist of western landscape and Native American life, served as the primary artist for the survey. Assisting him in the preparation of the expedition's artwork was a young German immigrant, Gustavus Sohon, who continued to record the region's landscapes and native population while accompanying Stevens on his 1855 treaty-making expedition and Mullan on his road-building project.

Isaac Ingalls Stevens

Isaac Ingalls Stevens was born in North Andover, Massachusetts, on March 25, 1818. In 1839 he graduated from the U.S. Military Academy at West Point. As a second lieutenant, Stevens began his military service with an assignment to the Corps of Engineers. During the war with Mexico (1846–48), while serving on the staff of Gen. Winfield Scott, Stevens was seriously wounded at the Battle for Mexico City. He never totally recovered from his injuries, and they plagued him the rest of his life. Returning to Washington, D.C., Stevens directed the U.S. Coast Survey Office from 1849 to 1853.

Stevens was an ambitious man. Biographer Kent D. Richards characterized Stevens as "a young man in a hurry." His ambition led him to resign his military commission in 1853 and accept two new assignments simultaneously—commander of the Pacific Railroad Survey's northern route and the first governor of the newly established Washington Territory, which also carried the responsibility of Superintendent of Indian Affairs for the region. He completed the railroad survey in 1853 as he traveled west to Olympia, Washington. In 1855,

Portrait of Isaac Stevens
—From Hazard Stevens, *The Life of Isaac Ingalls Stevens*

while serving as territorial governor, he conducted three major Indian treaty negotiations. His treaty-making expedition began in the Puget Sound region and extended to a site at the mouth of the Judith River, in central Montana.

During the same period, Stevens wrote the preliminary account of his railroad survey, which was published in 1855. It contained many interesting reports but did not include any significant illustrations. Subsequently, he prepared a final report, published in two volumes by the U.S. Congress in 1860. This report featured seventy plates (reproduced in this book's Gallery of Views), numerous maps, and a number of tables.

In 1857 Stevens was elected as the territory's representative to the U.S. Congress. He returned to Washington, D.C., and served in this capacity for both the 35th and 36th Congressional sessions. With the outbreak of the Civil War, Stevens resumed military service as a colonel in the 79th New York Highlanders. He was promoted rapidly to brigadier general. On September 1, 1862, after the Second Battle of Manassas, or Bull Run, in northern Virginia, he was killed during a severe thunderstorm while protecting Union forces as they retreated toward Washington, D.C. The honorary rank of major general was bestowed upon Stevens after his death.

John Mullan

John Mullan was a dreamer. He was born in Norfolk, Virginia, on July 31, 1831, but the family soon moved to Annapolis, Maryland. In 1842 Mullan enrolled at St. John's Academy in Annapolis, graduating with a B.A. degree in liberal arts. Seeking further education, he received an at-large appointment in 1848 to the U.S. Military Academy at West Point through the direct recommendation of President James K. Polk. Finishing fifteenth in his class in 1852, he began service at Fort Columbus, New York, as brevet second lieutenant.

Mullan eventually transferred to the Topographical Engineers. With his assignment to Stevens's Pacific Railroad Survey, he became intimately involved with transportation development and settlement promotion in the northern Rocky Mountains and the Pacific Northwest. He played a major role in exploring much of the land along the proposed route for the railroad. Especially significant were his forays into the mountains and valleys of present-day western Montana. One

Figure 3: Sohon's sketch of a section of the Mullan Road, in present-day Montana —Courtesy of National Archives and Records Administration

of his expeditions led to the discovery of Mullan Pass, west of today's Helena, Montana, and Isaac Stevens credited him with the discovery of the Little Blackfoot River.

Another task Stevens assigned to Mullan was the responsibility for collecting climatic data relating to the region's winters. Along with a small company of men, Mullan set up a camp during the winter of 1853–54 at Cantonment Stevens, adjacent to the Bitterroot River (then called St. Mary's River), some thirty miles south of current Missoula, Montana. He named the camp for Isaac Stevens, reflecting his admiration for his supervisor.

As the result of Mullan's work on the railroad survey in the northern Rockies and the Bitterroot Mountains, in particular, he was chosen in 1855 to begin a new project—the construction of a military wagon road between Old Fort Walla Walla, at current Wallula, Washington, to Fort Benton, in present-day Montana. Due to insufficient funding, the project was placed on hold. In the meantime, Mullan had several different assignments, including one in Florida to fight in the Seminole Indian War and another at a post in New Orleans. In 1858 he was assigned to renew the road project. This appointment was advocated by his old friend Isaac

Stevens, who in 1857 had become Washington Territory's delegate to Congress.

When Mullan arrived at Fort Dalles in mid-May 1858, however, he was unable to start the construction project because of Indian hostilities in eastern Washington. Consequently, he volunteered his services to Col. George Wright, who was attempting to bring the Indians to the treaty table. Mullan and his men were responsible for the topographical mapping on this expedition.

The following year, the War Department again ordered Mullan to begin construction of the road, and he was finally able to do so. The work began at the new site of Fort Walla Walla, and it was eventually completed to Fort Benton, for a total road distance of 625 miles. The construction lasted three years. Although scholars have criticized the construction quality of the road—now known as the Mullan Road—it was thoroughly planned and carefully built. The road's character is documented in a pencil sketch by Gustavus Sohon, who accompanied Mullan on this project (see figure 3). Although the road soon deteriorated, primarily due to lack of repairs during the Civil War, it was geographically and historically significant because it facilitated settlement in the eastern half of Washington Territory—present-day Idaho and western Montana.

Mullan eventually settled on a ranch near Walla Walla, Washington. From this location he continued to promote the region's settlement. Among other things, he wrote a travelers' guide entitled *Miners' and Travelers' Guide to Oregon, Washington, Idaho, Montana, Wyoming, and Colorado* (New York: William W. Franklin, 1865). For the remainder of his career, the retired army captain practiced law, first in Portland, Oregon, then in San Francisco, and finally in Washington, D.C. He specialized in land law. Mullan died December 29, 1909, at the age of seventy-nine. His body was taken for burial to Annapolis and today lies in an unmarked grave along with other members of his family.

John Mix Stanley

John Mix Stanley, recognized as one of the great mid-nineteenth-century artists of the American West, was the most prolific of the eleven artists accompanying the Pacific Railroad Surveys. He was responsible for the greatest number of plates in all the published reports. His appointment to Isaac Stevens's survey party was based upon his previous work in the West and the broad range of his artistic work—landscapes, portraits, and renderings of Native American life.

Stanley was born in 1814 in the town of Canandaigua, in the Finger Lakes region of upstate New York. Orphaned at age twelve, he was apprenticed to a coach maker two years later. There is no evidence that he had any formal training as an artist at this time, but by 1834 he had moved to Detroit, where he worked as a painter of signs and portraits.

Stanley's serious attention to western themes began in the 1840s, following his introduction to frontier life while at Fort Snelling, Minnesota, in 1839. Three years later, he traveled to Fort Gibson, in present-day Oklahoma, where he maintained a studio until 1845. During this interval, he painted portraits of Native Americans, which provided the basis for an exhibit that opened in Cincinnati in 1846. Many of these

Portrait of John Mix Stanley —From David I. Bushnell, "John Mix Stanley, Artist-Explorer," *Smithsonian Institution, Annual Report, 1924*

paintings became part of an exhibit displayed in Washington, D.C., at the Smithsonian's Gallery of Indian Art in the 1850s and early 1860s. Tragically, most of these (over two hundred) were destroyed in a fire in 1865.

The content and quality of the Smithsonian exhibit was most likely a key factor in Stanley's selection as the survey's main artist. The exhibit was widely acclaimed, and it is very probable that Stevens visited it while he was working in the Coast and Geodetic Survey Office in Washington, D.C. An added consideration in Stanley's selection was his extensive travel experience in the West.

Of the seventy landscape illustrations in Stevens's final report, fifty-five were attributed to Stanley. In addition, he redrew three or more landscapes originally drawn by others. When Stevens's survey was initiated in Minnesota on May 27, 1853, Stanley was assisted by another artist, Max Strobel. But Strobel could not tolerate the pace of the expedition and the amount of work required, so he left the survey party within a few days. It appears that Stevens and Stanley were both workaholics—few others had the ability to keep up.

The expedition moved rapidly. By September 22, the group had reached Fort Benton, Nebraska Territory. While camping there, the party was joined by personnel who had traveled from the West Coast—the most notable addition being a young German immigrant named Gustavus Sohon. Stanley and Sohon soon established a friendship based on their mutual interest in art. Consequently, Stanley trained Sohon as the survey's backup artist.

Most of the lithographic views in the published reports were based on small, colorful watercolors made in the field. Many of these watercolors still survive and are housed at the Yale University Art Gallery. For the readers' appreciation, several samples of these beautiful illustrations are included in Part 3 of this book. One example appears in this section, "Butte de Morale," plate 13.

Based on the published record, we know that Stanley sketched numerous landscape scenes as the Stevens survey party moved westward. However, he did not visit all of the sites attributed to him, so it is obvious that others played a role in providing materials, namely Sohon and Dr. James G. Cooper, one of the scientists working in the Puget Sound region. They probably prepared simple pencil sketches that Stanley later rendered as full watercolor images.

Stanley was interested in new technologies. During the 1840s, he learned how to make daguerreotypes, an early form of photography. He used photographic equipment on Stevens's survey and is known to have taken Native American portraits at Forts Union and Benton in midsummer 1853. None of these images are known to have survived; they were probably destroyed in the Smithsonian fire in 1865.

Upon the completion of the expedition in 1854, Stanley was sent east with the preliminary reports. There he finished the illustrations to be used in the published reports. He received $125 a month for this task.[5] Crop marks and other pencil notations were added to the original watercolors to indicate specific details that Stevens and Stanley wanted to change or delete in the final lithographic prints.

Stanley remained in Washington, D.C., until 1863. He then moved to Buffalo, New York, and shortly thereafter to Detroit, Michigan. When the 1865 fire destroyed 50 percent of his western art, the artist was devastated. He attempted to re-create the originals, but he failed at this task and became despondent about the loss of his life's work. This emotional state probably contributed to his early death at the age of fifty-eight.

Stanley's career had begun in Detroit, and it ended there with his death in 1872. During his lifetime, he created over four hundred paintings. Of these, approximately two hundred still exist. He also painted two large panoramas—one illustrating sixty-seven Civil War battle scenes, the other portraying forty-two scenes drawn from his western travels. Unfortunately, neither of these has survived. The Fine Art Dealers Association Web site summarized Stanley's life work:

> John Mix Stanley is best known today for vivid documentation of American Indian life in the West before it was swept away by advancing settlement. . . . He experimented between fine and popular art traditions, adapting his artistic language to the medium or patron for which he was painting. He was greatly influenced by both contemporary American history painting and early nineteenth-century European art. He was among the first artists to document topography of the West and be allowed to paint portraits of various Indian tribes.[6]

Gustavus Sohon

One of the very talented, though little recognized, men of the Stevens expedition was Gustavus Sohon. This humane

Portrait of Gustavus Sohon —Courtesy of Sohon family

and modest man was born on December 10, 1825, in Tilsit, Prussia (today's Sovetsk, in the Kaliningrad province of Russia). To avoid service in the Prussian Army, he emigrated to the United States. Interestingly, at the age of twenty-seven, Sohon enlisted in the U.S. Army. His original unit, Company K, 4th Infantry, was destined for duty in the Pacific Northwest. He joined Stevens's survey party in September 1853.

Sohon is now identified as a very important artist of the mid-nineteenth-century American West. His major contributions occurred during a nine-year period, approximately 1854 to 1863, when he worked on the Pacific Railroad Surveys and the Mullan Military Road. While a number of Sohon's original drawings have been published in secondary works during the last fifty years, his works were initially seen in published accounts of expeditions. Specifically, ten of his drawings were included in Stevens's Pacific Railroad report, while other Sohon drawings were published in the final Congressional report for the Mullan Military Road in 1862, and in Hazard Stevens's biography of his father, published in 1900.[7]

Aside from his artistic talents, Sohon was noted for his linguistic abilities. He spoke three languages—German, French, and English. In addition to these, Sohon rapidly acquired the nuances of the Salishan language spoken by the Flathead and Pend d'Oreille Indians. The combination of interpretative, artistic, and scientific skills enabled Sohon to play a pivotal role in both the Pacific Railroad Survey and the Mullan Road construction.

Following his appointment with the Stevens survey in 1853 and 1854, Sohon assisted Stevens in his 1855 treaty negotiations with Indians in the region. Sohon worked as one of the official interpreters for two key treaties—one negotiated near the Walla Walla River in Washington Territory in May 1855, and one at Missoula in July 1855. A third treaty was undertaken in October 1855 near the junction of the Missouri and Judith Rivers.

In 1857 Sohon began working as a civilian under John Mullan, who was constructing a military road connecting the Columbia and Missouri Rivers. Sohon was a major contributor to this project because he was a "jack of all trades." Sometimes he worked as a topographer, making maps. On other occasions he scouted out potential routes or passes for the road. In addition, Sohon frequently collected meteorological data and drew pictures of the road or nearby landmarks. Upon the completion of the Mullan Road, Sohon traveled east with Mullan, and in 1862 he was asked to help prepare the data for the final report to Congress.

The remainder of Sohon's life was puzzling. Instead of continuing to work as an artist, Sohon turned to a modest career as a shoe salesman in Washington, D.C. Sohon, the unsung explorer and artist of the Pacific Northwest, died September 3, 1903, at the age of seventy-eight.

Sohon did not keep a journal or retain much documentation about his own life. Thus, we know little about his background, his education, or how his artistic talents developed. However, considerable artwork survives. The authors have documented 219 pieces, which compose the primary evidence of what Sohon accomplished.[8] One-third of Sohon's drawings were of Native Americans, the vast majority being portraits. Since John Mix Stanley drew many similar images, one can assume that Stanley had a strong influence on Sohon's artwork in both subject matter and artistic style.

The noted western historian William H. Goetzmann summarized Sohon's work: "Private Sohon, a self-taught artist, had incredible talent that he never really developed as he confined himself to penciled Indian portraits and modest sketches."[9] Nevertheless, these "modest" sketches have considerable value. They provide us with some of the first detailed views of this mountainous landscape and its native peoples during the mid-nineteenth century.

The Published Reports

Congress originally allocated $145,000 for conducting the Pacific Railroad Surveys. Ironically, the government paid more to publish the results of these explorations—$1,000,000—than to conduct them. Even so, the investment was worthwhile, for it provided the nation with its first comprehensive geographic survey of the West. The surveys' reports were published in twelve volumes as House and Senate documents, now part of the Congressional Serial Set.

As his contribution to this twelve-volume compendium, Isaac Stevens prepared both a preliminary report, which was published in 1855 as part of volume 1, and a final report, which was first published in 1859 as a one-book supplement to volume 1, then slightly revised in 1860 as volume 12 (in two books).[10] Volume 12 was the most richly illustrated of all the survey reports. The first book featured Stevens's general report, providing the geographical narrative of the expedition. It contained the seventy landscape views we have reproduced in this book. The second book contained a botanical report, including a catalog of plants found east and west of the Rocky Mountains, and a thorough zoological investigation beginning with insects and extending through crustacea. These botanical and zoological reports included fifty plates illustrating the region's rich treasure of flora and fauna (see figure 4).

Stevens's general report in the first book of volume 12 was subdivided into eighteen chapters. The first ten chapters provided the narrative for the 1853 surveys, which were only summarized in the 1855 preliminary report. Each chapter discussed the geography of specific sections along the route, beginning with St. Paul and concluding with the Puget Sound. Chapters 9 and 10 recorded the secondary explorations of John Mullan and James Doty, also conducted in 1853 but not covered in the preliminary report. The next two chapters documented Stevens's findings during 1855 when he conducted his treaty expeditions between Puget Sound and Fort Benton, Montana.

Figure 4: "Western Duck Hawk," sample of zoological illustration —From Isaac I. Stevens, *Narrative and Final Report of Explorations for a Route for a Pacific Railroad,* vol. 12, bk. 2, plate 11

The final chapters include a "Geographical Memoir" for the region, providing a narrative about the region's mountains, hydrography, and meteorology. The section on hydrography was contributed by Johann G. Kohl, a German scholar who worked for the U.S. Coast Survey. The book concludes with a "Railroad Report and Estimate" and a discussion of how altitudes were computed from barometrical observations. There are also two appendices concerning heights and distances and meteorological observations, as well as a fold-out map, an isothermal chart, and a sheet of topographic profiles.[11]

The seventy landscape views were dispersed appropriately throughout the first twelve chapters. They generally represent a linear progression that originates at Camp Pierce, near Fort Snelling, and ends on the shorelines of Puget Sound near Whidbey Island. Because the 1860 publication also included reports created in 1854 and data collected when Stevens was conducting his treaty-making expedition in 1855, the image sequence breaks pure linear progression by incorporating some images in these auxiliary reports that appear out of place temporally and geographically. They were added arbitrarily to show places visited and events that occurred after the main party's work was finished in early winter of 1853.

The 1860 landscape views are lithographic reproductions. The artists most likely recorded their observations in the field as rough pencil sketches, but very few examples of the field drawings survive. From these sketches, they created final pencil and watercolor renditions. These original "manuscripts" were used to prepare the lithographic stones from which the final illustrations were printed. Each lithographic print represents an interesting integration of line work emphasized in black ink with tinted background in assorted colors. To create multicolored images, several stones were prepared, and the image was printed in stages. Typically, the published prints feature three colors representing four printing stages—a black plate plus one plate for each color.

In the published reports, the paper stock used for the lithographs was different than that employed for the text. It is whiter and thicker, and characterized by a smooth coated finish. The use of different paper stock suggests that the illustration plates were printed separately from the text and later integrated upon binding. Further evidence is that the illustrations are not always bound in the same place in duplicate volumes.

Two different printing firms were used in preparing the lithographs for the two printings of the final report. The prints in the 1859 report (the supplement to volume 1) were prepared as lithographs by Julius Bien, New York, while those in the 1860 report (volume 12) were printed by Sarony, Major & Knapp, also of New York City. Not surprisingly, there are subtle differences between the two productions. In the Bien collection, the images are soft and subdued. The Sarony, Major & Knapp prints, which are reproduced in the present work, were enhanced and thus appear more dramatic and colorful.

Creation of the Illustrations

Two artists—John Mix Stanley and Gustavus Sohon—created most of the images depicting scenes along the routes followed by the military explorers. Officially, Stanley created fifty-seven; Sohon, nine; and Dr. James G. Cooper, the expedition's surgeon and naturalist, contributed one. Stanley and Sohon collaborated on three images, while Stanley and Cooper collaborated on another two images. In these, Sohon or Cooper made the original sketch and Stanley completed the final artwork for the lithograph.

When examining the images used to illustrate the report, one intuitively asks how closely they represent what the explorers saw as they traveled through the region, or whether they depict real events that occurred on the expedition. Who actually created the original artwork? Was the illustration composed in the field, or is it a reconstruction? The answers are not always available. Some questions have been answered upon investigation, but others remain unresolved.

Certainly, by looking at each image carefully, one garners valuable insights. This insight is sometimes complemented by narrative contained in the report, or perhaps by other primary or secondary documentation. In trying to understand the works of Stanley and Sohon, more information is available for Stanley, since Sohon kept little documentation.

Another intriguing consideration regarding the creation of these images concerns photography. It has been documented that Stanley used daguerreotypes at Fort Benton and Fort Union. Some scholars surmise that a few scenes in the Pacific Railroad reports may have originated as photographs, and that these were the source materials for the lithographic prints. Fort Benton is a case in point (see figure 5). In addition, Sohon later took up this interest and for several years ran an ambrotype studio in San Francisco (1864–66).

In general, the authorship of the seventy images contained in Stevens's report is faithfully identified. However, there are a number where the authorship is suspect. It is known that Stanley did not visit some of the sites for which he is given credit. We know Stanley never visited Fort Okinakane or Whidbey Island, for instance. There is little likelihood that he ever visited Lolo Hot Springs and Flathead Lake, yet he is credited with preparing these drawings. Rather, these landscape views were probably based on sketches made by Sohon. The three images depicting Puget Sound landscapes originated from sketches contributed by Dr. Cooper, the expedition's naturalist.

The problem becomes more complicated when dates are ascribed to the images. Stanley's image of Peluse (Palouse) Falls was probably not drawn in the fall of 1853, when he accompanied Stevens south toward Old Fort Walla Walla. He did not have enough time to visit this site. Also, he did not

Figure 5: "Fort Benton" —From Isaac I. Stevens, *Narrative and Final Report of Explorations for a Route for a Pacific Railroad,* vol. 12, bk. 1, plate 24

Figure 6: "Cantonment Stevens, Looking Westward" —From Isaac I. Stevens, *Narrative and Final Report of Explorations for a Route for a Pacific Railroad,* vol. 12, bk. 1, plate 59

visit the Grand Coulee during this time, yet his authorship is placed on this image. Perhaps some of the mystery can be clarified by taking into account Stanley's visit to the Oregon Country six or seven years earlier. Maybe previous artwork was incorporated into the report to fill in gaps. Another possibility is that someone else drew the original sketch but was never properly credited for it.

One might assume that the printed illustrations represented all the drawings created on the expedition, but this supposition is incorrect. Stanley created many small drawings that were not included in the final report. A few of these are now in the Detroit Institute of Art, including one sketched near Fort Union showing tribal activity and another done along the upper Milk River. Both were drawn in the summer of 1853.

Images contained in the report were printed in one of three formats—landscape, portrait, or oval. Most were done in a landscape, or horizontal rectangular, mode. Two images employed a portrait, or vertical, format. Both were waterfalls. Several were composed as ovals; all of these were done by Sohon, who used an oval to define the neat line of the image, such as in his rendering of Cantonment Stevens (see figure 6). The use of an oval for defining the image is one way to distinguish images drawn by Sohon from those created by Stanley.

Artistic Techniques and Composition

As one examines the seventy images used to illustrate Stevens's report, several questions arise. What was Stevens purpose in selecting these images, and how factual were they? Immediately, one concludes that each was intended to document what was seen along the route. This is a valid observation. However, Stevens had several other goals. First, he wanted this route to be selected for the nation's first transcontinental railroad. Second, he wanted to promote the settlement of Pacific Northwest, which was then known as the Oregon Country. To accomplish these objectives, he needed to present a positive viewpoint to Congress and other readers of the report. Consequently, many of the images portray picturesque and peaceful settings. Reality was often changed or embellished through artistic techniques and composition to accentuate these promotional objectives.

Although it appears that Stanley and Sohon had little formal training as artists, analysis of the published landscape views reveals that both followed established composition rules and artistic techniques. For example, they employed the rule of thirds. In its application, an illustration is visually divided into thirds, both horizontally and vertically. Sometimes in preparing a drawing, the artist may have actually drawn lines on the paper or canvas as a preliminary step. The artist then placed a subject at one of the intersection points, which are power points. (See figure 7.) Preliminary analysis of the images reveals that Stanley, a more experienced artist, used power points more often than Sohon did.

Another artistic device used in composition is the placement of human figures somewhere within the scene to provide a sense of scale. Landscape magnitude is defined by comparing man to the region's physical geography. Man is small in relationship to the natural landscape.

The only panoramic view included in this collection, "Main Chain of the Rocky Mountains," demonstrates very well the sense of scale and the magnitude of these mountainous landscapes (see figure 8). As the largest illustration in the report, it presents an elongated view of the Rocky Mountain Front as viewed from the east, somewhere in the vicinity of today's Choteau, Montana. In the background, a diminutive outline of the mountain range extends from the south, on the left side of the image, to the north, on the right. To aid identification, a few key mountains and important passes are labeled by name. The foreground is dominated by a Native American aiming his rifle at a small herd of antelope that are peacefully grazing in a depression below. Death is near. Midground reveals a line of men moving with equipment and animals through the scene.

Besides providing a sense of scale, the use of human figures accomplished several other objectives. First, they documented the presence of people—a once wild landscape is now being opened to development. Second, they represented an attitude toward this development. Perhaps a railroad will soon be constructed and the land will become, for the first time, civilized. Now landscape has a use.

Another technique that is evident in these views is the placement of plants dominantly within the scene—especially if the plant was unique or unusual. Perhaps the artist had

Figure 7: Rule of thirds applied to Cottonwood Grove, where a cottonwood tree is positioned at the intersection of two lines —From Isaac I. Stevens, *Narrative and Final Report of Explorations for a Route for a Pacific Railroad,* vol. 12, bk. 1, plate 19

Figure 8: "Main Chain of the Rocky Mountains" —From Isaac I. Stevens, *Narrative and Final Report of Explorations for a Route for a Pacific Railroad,* vol. 12, bk. 1, plate 52

Figure 9: "Butte de Morale"
—From Isaac I. Stevens, *Narrative and Final Report of Explorations for a Route for a Pacific Railroad,* vol. 12, bk. 1, plate 13

never seen it before; this alone made it worthy of inclusion. The plate showing Butte de Morale is an example (see figure 9). Stanley was captivated by the gnarled and twisting tree limbs along the butte's ridgeline. The trees show the ravages of time and continuous weathering. They also suggest a wilderness untouched by the human spirit.

Water features are a landscape element in many of the images. As the explorers moved westward, they followed river courses and encountered many lakes. They often rendered water bodies as placid. But is the water sometimes too quiet to be realistic? Near Fort Benton, Stanley drew a scene showing a meandering Missouri River glistening in front of the trading post (see figure 5). The river's surface is strangely smooth. Some have suggested that this is the result of Stanley's using a daguerreotype image as his original source, in which the camera's long exposure time would have blurred or smoothed the appearance of the river's movement. But perhaps the effect is an intentional alteration of reality. The peaceful appearance of the river is accentuated by the placement of horses quietly drinking as several Native Americans watch over them. This interpretation suggests that the native peoples are at peace with nature. There is a harmony between them, the animals, and the surrounding landscape.

On the other hand, the depiction of turbulent waters flowing down the Columbia River, west of the Rockies, provided a different message. Two examples are Stanley's images of Kettle Falls (see figure 10) and the rapids at the mouth of the Deschutes (see figure 11). In both cases, cold, clear water cascades over the rocks along the river's channel. The visual symbolism of these dramatic scenes is nature uncontrolled, power wasted. At least, the river could be exploited for its salmon and trout.

In examining the image showing Puget Sound from Whidbey Island (see figure 12), one may observe a different perspective, relating to the hazards and impediments turbulent waters presented to travel and commercial activity. Here the water's surface depicts violence, agitated by the combined effects of wind and currents. In the foreground, people in

Figure 10: "Kettle Falls, Columbia River" —From Isaac I. Stevens, *Narrative and Final Report of Explorations for a Route for a Pacific Railroad,* vol. 12, bk. 1, plate 47

Figure 11: "Columbia River, Junction of Des Chutes" —From Isaac I. Stevens, *Narrative and Final Report of Explorations for a Route for a Pacific Railroad,* vol. 12, bk. 1, plate 67

Figure 12: "Puget Sound & Mt. Rainier from Whitby's [sic] Island" —From Isaac I. Stevens, *Narrative and Final Report of Explorations for a Route for a Pacific Railroad,* vol. 12, bk. 1, plate 68

a large dugout canoe struggle to navigate the boat safely. The implied message suggests that man is now confronting nature directly. Similarly, the construction of a railroad will face confrontations with nature, but progress will slowly be made and the obstacles in its way will be overcome.

Finally, the interplay of light and dark is another artistic technique that is very evident in the printed illustrations. Many of the views display grandiose, cloud-filled skies penetrated by rays of sunlight. Few show an open or "bald" sky, which in photographs and artwork is generally thought to be uninteresting and is therefore avoided. However, this artistic element may have been added during lithographic printing, because the artist's original watercolors are characterized by attractive landscapes but bland skies.

Light interacting with clouds and landscape does provide a sense of drama within the typical scene. It also creates image depth—an impression of three dimensions. On a metaphorical level, the sun projecting its rays upon the land symbolizes the presence of God, perhaps suggesting that it was destiny that a railroad should be constructed through this landscape.

Conclusion

The illustrations accompanying the report of Isaac Stevens's Pacific Railroad Survey are especially noteworthy. Not only do these drawings depict the first published illustrations of many places along the proposed route, they also provide interesting, visual insights about the cultures of peoples found along the route and the day-by-day activities of the survey party.

Each view depicts a landscape in the northwestern United States, either undisturbed wilderness or places in the initial throes of development, including missions, forts, or rudimentary towns. These drawings provide a visible record of what was seen and perceived important by their creators, especially Stanley and Sohon. The publication of these early views of the Pacific Northwest in the final reports of the Pacific Railroad Surveys gives us visual documentation of the early life of Native Americans and of the grandeur, magnitude, and solitude of the nineteenth-century American West.

Part 2

The Images Tell a Story

THE SEVENTY LANDSCAPE VIEWS created specifically for Isaac Stevens's final congressional report, published in 1860, are reproduced as full-size color plates in the Gallery of Views that follows. Together they document the landscape and people Stevens and his men encountered along the route of the Northern Pacific Railroad Survey, which took place between May 1853 and September 1854.

The visual journey we have re-created here is enlightened by excerpts from the report's narrative and by historical commentary. A few plates are accompanied by the original watercolor sketches so the reader can see how closely the lithographic prints resemble the original source material. In addition, we place the illustrated locales in a modern context by including many modern photographs and location maps, as well as information about the site's geography and geology.

Modern Maps

We have created two sets of maps, which are designed to provide geographical information relating to the illustrated sites. The first set of six sectional index maps are small-scale maps delineating major portions of the route traveled by Stevens's party, superimposed on portions of a historic map originally compiled by Lt. Gov. G. K. Warren.[12] First published in 1857, Warren's map was the most accurate depiction of the West's geography as known in the late 1850s. Although it was a composite map, incorporating information gained from prior explorations, it also depicted much of the knowledge acquired during the Pacific Railroad Surveys, including that of Stevens's survey. Each index map portrays about 400 linear miles of landscape.

A second set of larger-scale maps accompany the landscape views as reproduced in the Gallery of Views. They are designed to indicate the immediate region around the illustration site. While most of the sites the survey party visited are identified in specific geographic terms, a few are approximations of where the image was rendered. For example, it is not known precisely where Stevens visited the Assiniboine people on the northern plains. River crossings are also difficult to identify. Sometimes the explorers crossed a river several times. The main party crossed the Bitterroot River numerous times in October 1853, but the particular crossing shown in Stanley's drawing is not known. Only the general location—somewhere between Alberton and St. Regis, Montana—can be ascertained.

In order for the reader to understand the setting of each site, the maps have been drawn in a color, shaded-relief format. The shaded relief was developed using digital elevation model (DEM) data supplied by the U.S. Geological Survey. All the maps are oriented to the north.

Most lakes, rivers, and streams along the exploration route are depicted on these maps. Bear in mind, however, that hydrographic patterns of the mid-nineteenth century are not the same as those of today—many rivers have changed their course with the passage of time or have been flooded with the construction of dams. Outlines of former river courses are shown on the maps with a blue dashed line.

The place-names, including river names, used on the maps are the modern ones. Place-names commonly change, often reflecting geographical, political, or cultural changes. Several of the places identified by Stevens have since been renamed, such as some of the rivers near Missoula, Montana. In 1853, portions of the Clark Fork had entirely different names. Upstream from today's Missoula, it was called the Hell Gate, and downstream toward the northwest, it was the Bitter Root. What is today the Bitterroot River, south of Missoula, was once called the St. Mary's.

To aid the reader, important historical landmarks are identified on the maps. For example, the trading posts at Old Fort Walla Walla and Kettle Falls, the sites of which are now submerged under reservoirs, are marked where they existed. Some trading posts, such as Forts Union, Vancouver, and Benton, have been either restored or reconstructed, but not necessarily on their original site. On the maps, we show their historical locations. Significant modern landmarks are also included on the maps where appropriate, as are present towns and cities, and in some cases, cultural features such as wildlife refuges, state parks, and military bases. In most cases, modern transportation networks and selected railroads are included on the maps. In urban locations, grids are symbolic rather than accurate representations of the actual streets.

One of the authors' goals in creating this book was to establish, if possible, the locations where John Stanley and Gustavus Sohon created their images. On the larger-scale maps, circular number symbols contain the plate numbers for specific images. In many cases, a small triangle appears adjacent to the number; this symbol approximates the artist's viewpoint.

Arrangement of the Plates

The seventy landscape views, which were printed as chromolithographic plates, are arranged in the Gallery of Views as they were in Stevens's final published report. They were numbered from one to seventy, using Roman numerals. What may cause some confusion is the fact that some of the plates do not run in a linear east-to-west progression. The plates approximate a geographic line from St. Paul, Minnesota, to Whidbey Island, Washington, but the sequence becomes jumbled from the Rocky Mountains westward.

The first break in the east-to-west sequence comes at plate 44 (Fort Vancouver). This image was created in 1854 after the main survey was completed. It is out of place geographically and temporally relative to the next three plates, which depict the Cascades of the Columbia (plate 45), Cape Horn (plate 46), and Kettle Falls (plate 47), all east of Fort Vancouver. Stanley produced those three images in 1853.

The next break in the geographical sequence was the result of Stevens's team later adding reports of work completed by other parties—especially the one headed by Lt. John Mullan. From October 1853 to August 1854, Mullan explored the Rocky Mountain region from the Canadian border south to Fort Hall in Idaho. Plates 46 through 65 document his intermontane explorations.

The final five plates include two additional plates by Stanley (plate 66, Bitterroot River near Fort Owen, dated October 1853, and plate 67, Columbia Junction of the Deschutes, dated November 1853). Views of Mount Rainier, Puget Sound, and Mount Baker conclude the report (plates 68, 69, and 70). These were probably sketched late in 1853 or in the spring of 1854. Authorship for these images is attributed to Stanley and Dr. James G. Cooper as co-artist.

Overview of the Expedition's Trek

In sixteen months, Stevens and his men explored a considerable geographic area, traveling more miles and covering more territory than Lewis and Clark did fifty years earlier. A conservative estimate for the distance traversed is 11,000 miles. What is even more amazing is the fact that much of the area was explored in parties of one. Each was vulnerable to attack or accident with no one on hand to help.

The expedition was well planned and well organized. Broken into two major sections, the work to explore a possible railroad route was subdivided between a western column, headed by Capt. George B. McClellan, and an eastern group, headed by Isaac Stevens. McClellan explored the Cascade Mountain region into the Columbia River Plateau, and Stevens explored the cross-country route from St. Paul, Minnesota, to Puget Sound. One of the main goals of the surveys was to find passes through both the Rocky and Cascade Mountains suitable for a railroad. Another major goal was to document winter conditions in the mountains, especially snowfall.

To assure the provision of supplies for the expedition, Stevens assigned Lt. Rufus Saxton to assemble a supply column at Columbia Barracks, adjacent to Fort Vancouver on the Columbia River. From there, Saxon headed east to a depot at Fort Owen (Montana), in the Bitterroot River valley east of the Bitterroot Mountains. Another small party, including Lt. John Mullan, took a steamboat up the Missouri River to Fort Union. Its task was to ascertain the feasibility of using steamboats to transport supplies for railroad construction.

Stevens's component of the expedition began at St. Paul, Minnesota, a newly created settlement on the northeast bank of the Mississippi River (see index map 1). He established his initial camp within a few miles of Minnehaha Falls. From there he proceeded by steamboat upstream on the Mississippi. Upon reaching the Saux River, he disembarked and began a long overland trek westward on foot. The journey lasted six months and covered about 7,000 miles. In addition, side expeditions covered additional territory, collecting data specifically to establish possible alternative routes and climatic conditions in areas adjacent to the general route.

Stevens's party crossed lands that would become the north-central and northwestern states of the United States. A portion of the journey followed the course explored by Lewis and Clark in 1804–6. It was the first major investigation of the territory since then. Most of Stevens's path paralleled rivers or streams. He followed these hydrographic features because they generally define a reasonable gradient that could be exploited by railroad builders. They usually wanted the gradient to be less than 2 percent per mile.

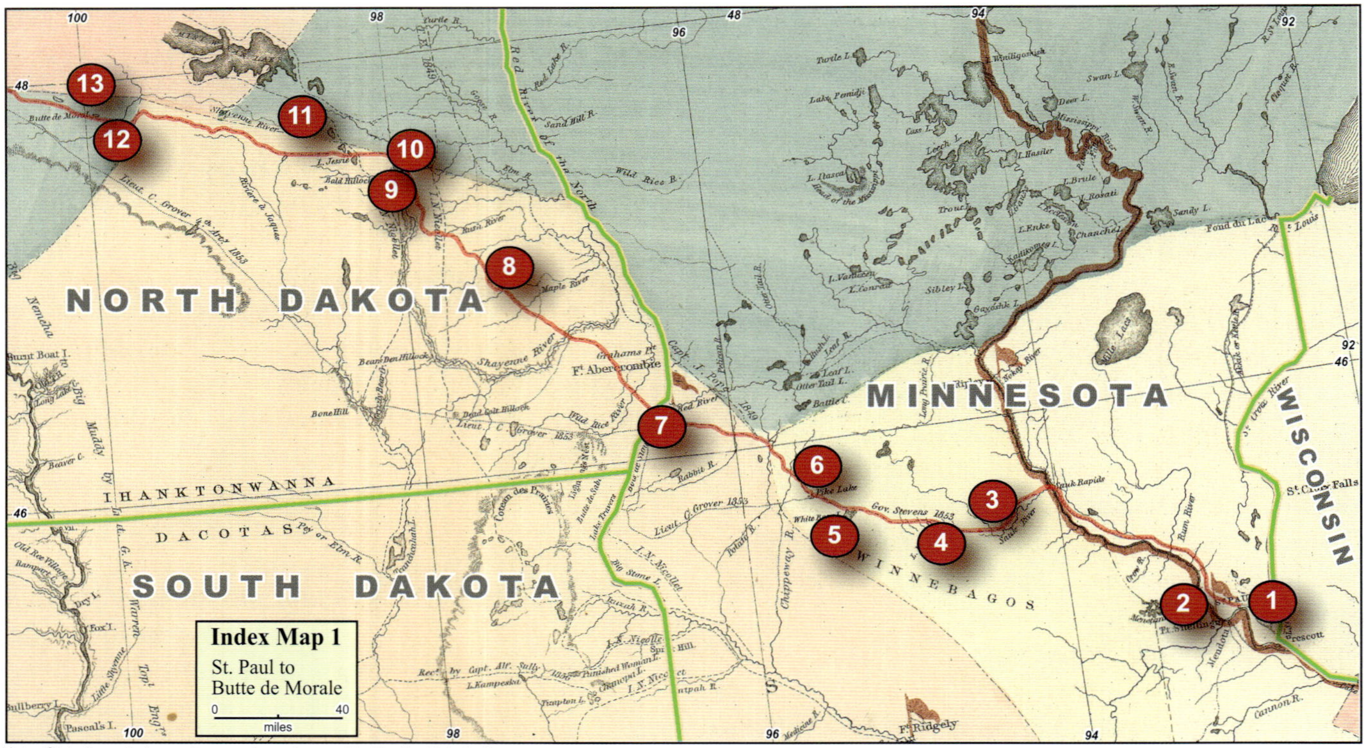

Note: State boundaries and names have been added to the historical map.

The journey proceeded through central Minnesota. Along this section, the expedition encountered numerous glacial lakes, both large and small. These are known by geologists as "kettle lakes." They were the product of continental glaciations and the retreat of ice as the Pleistocene Age ended about 10,000 years ago. As the party entered the Great Plains, the tree-dotted landscape yielded to prairie grasses. In the Dakotas, the men saw buffalo in large numbers. They were visually overwhelmed when they encountered a herd estimated at 250,000 near Lake Jessie.

During the first few months of the survey, Stevens and his party were very observant. Stevens made a considerable effort to note cultural details about the native peoples they encountered, but as the months dragged on, he became impatient to reach his destination. Consequently, information regarding different native groups diminished over time in both quality and quality.

One of the survey party's first encounters with people on the plains was with a group Stevens identified as the Red River Hunters. This large band made twice-yearly buffalo-hunting trips for meat and hides. Many of their buffalo robes were eventually traded at Fort Snelling and St. Paul. Another group they visited were the Assiniboines, whom they met just east of Fort Union (see index map 2). Stevens counted 150 lodges housing about 1,500 people. At one time the tribe was larger, but it had been ravaged by smallpox years earlier.

Upon reaching Fort Union, a major trading post on the north bank of the Missouri River, the party replenished some of its livestock and food supplies. After a brief stay, the group followed the Missouri River westward along its northern bank. Several days later, the party reached the river's junction with the Milk River, then followed the Milk River corridor westward until reaching the approximate location of Havre, Montana, in mid-August. While in this area, the explorers were treated to a spectacular night sky dominated by Comet Klinkenfus for several weeks. Interestingly, the comet was also observed by Capt. George B. McClellan during his corresponding surveys in the Puget

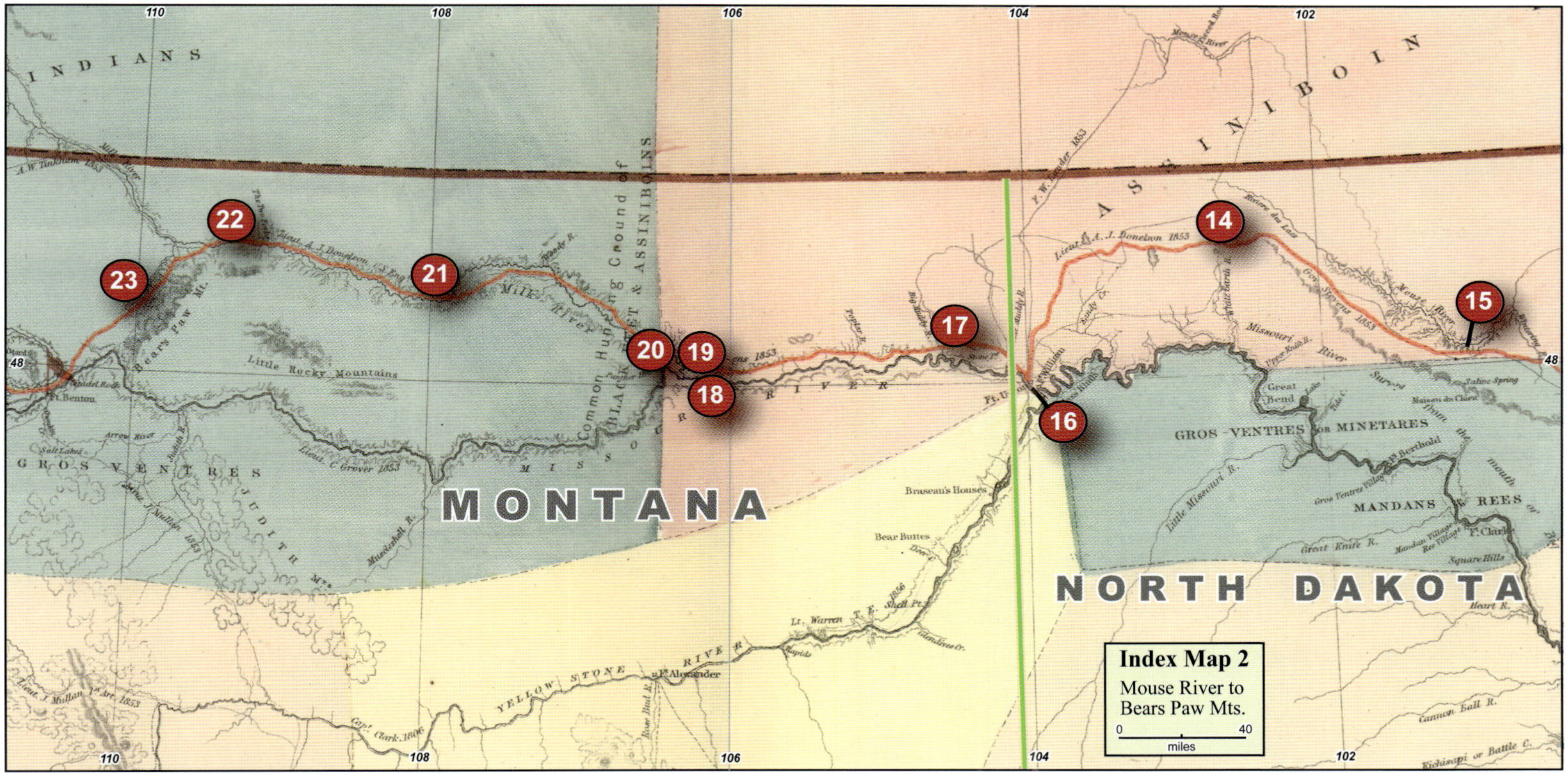

Note: State boundaries and names have been added to the historical map.

Sound region. He provided more scientific commentary on the comet than Stevens did.

The expedition turned southwest toward Fort Benton, another major trading post on the Missouri. Along the route, a small party explored the Bears Paw Mountains to the east, and while the main party was encamped at Fort Benton, another small group explored the Three Buttes region to the north-northwest (see index map 3).

Stevens left Fort Benton in mid-September, with his main party moving southwest. They traveled through a land characterized by spectacular flat-topped buttes interjected with numerous igneous intrusions. Late in September they encountered the Dearborn River, which they followed into the heart of the Rocky Mountains. They found the Continental Divide far easier to cross than they had expected. Within a small geographical area, three mountain passes were possible choices; Stevens chose Cadotte Pass. Upon crossing the divide, the group descended along the western slopes to a corridor formed by the Big Blackfoot River—the path of an Indian trail.

Encountering the junction of the Big Blackfoot and Hell Gate Rivers, the party moved northwest along the Hell Gate, today called the Clark Fork (see index map 4). Near present Missoula, Montana, they observed a huge elongated valley—originally called the Hell Gate Ronde. It is a curious feature because it lacks significant tree cover and is covered by luxuriant grasses. Along the side of the valley, the men observed hundreds of parallel terraces that marked the various shorelines of what was once a huge glacial lake, referred to as Lake Missoula. This landform complex was created by the repeated filling and draining of the lake, which occurred as the advance of a continental glacier created numerous dams over time. When the ice dams were breached, huge floods surged downstream, creating spectacular landforms in eastern Washington state—these are still referred to as the "scablands."

After a two-week visit at Fort Owen in the Bitterroot Valley, Stevens moved the main party westward again. Winter was approaching, and the governor was anxious to reach the

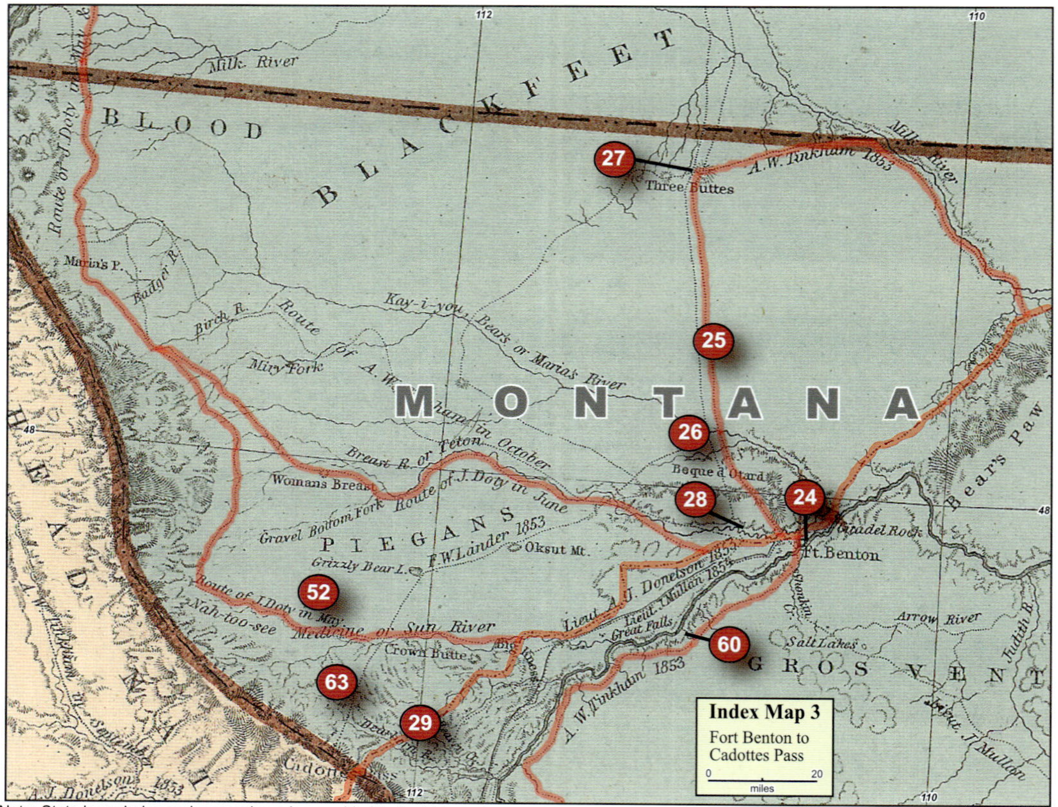

Note: State boundaries and names have been added to the historical map.

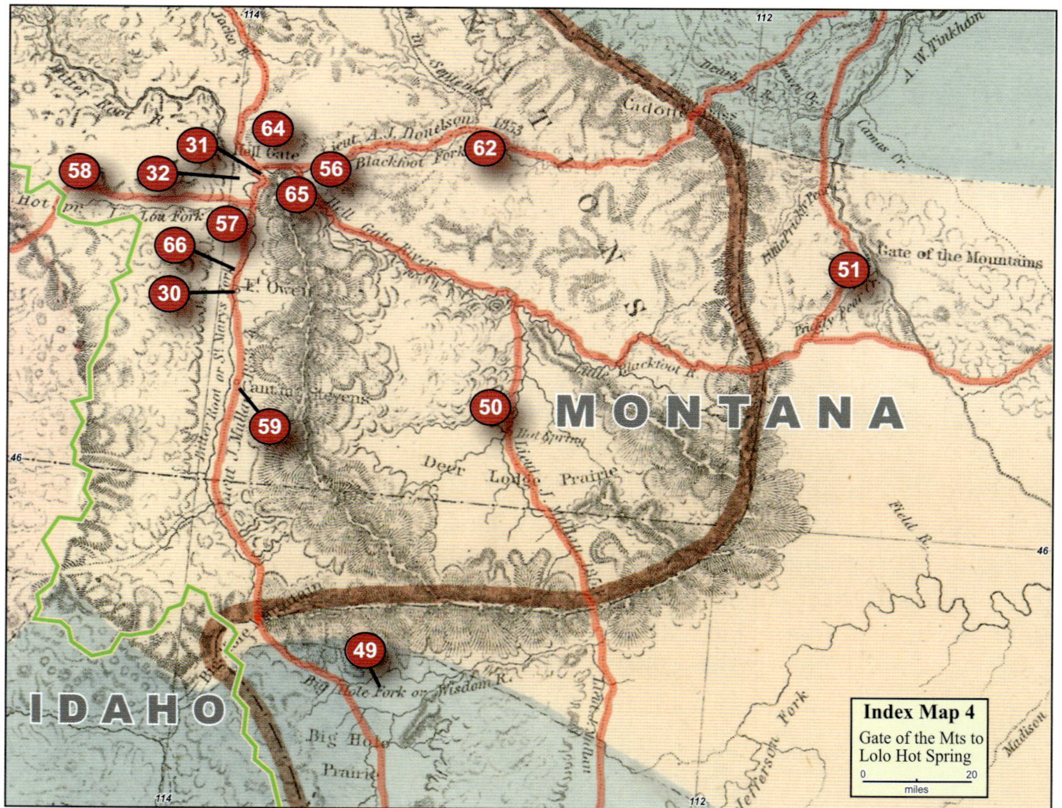

Note: State boundaries and names have been added to the historical map.

new territorial capital at Olympia. But he was aware of two major shortfalls in his exploration for a railroad route. First, he lacked meteorological data for the region, especially regarding winter snowfall. Second, he did not know what other routes could be found through the Rockies and Bitterroots. To address these questions, he left behind parties to collect data and conduct further explorations. One party was led by Lt. John Mullan. He set up his base of operations on St. Mary's (Bitterroot) River, naming the camp Cantonment Stevens.

In mid-October, Stevens moved northwest along the Hellgate River (see index map 5). The main party followed the broad valley formed by Glacial Lake Missoula, but near Alberton, Montana, they entered a more confined structure defined by large pine forest, mountain spurs, and many river crossings. Upon reaching the mouth of the St. Regis River, the group proceeded up the narrow St. Regis toward the crest of the Bitterroots. It was difficult going because the valley was narrow, rugged, and filled with trees and brush. Crossing the Bitterroots was a formidable task—these mountains proved to be a bigger obstacle than the Rockies.

After resting at the Coeur d'Alene Mission, also called Sacred Heart Mission or Cataldo Mission, on the banks of Idaho's Coeur d'Alene River, Stevens and his party resumed their travel westward along the Coeur d'Alene until its junction with the Spokane River. The men moved downstream through a broad river valley, passing a thunderous waterfall near the site of today's Spokane, Washington. The party stayed briefly at a trading post along the banks of the Spokane, then proceeded directly south across the Columbia River Plateau, a barren landscape characterized by ancient lava flows and coulees. Floodwaters from ancient Lake Missoula produced a vast scabland consisting of braided channels etched into the hardened surface of former lava flows.

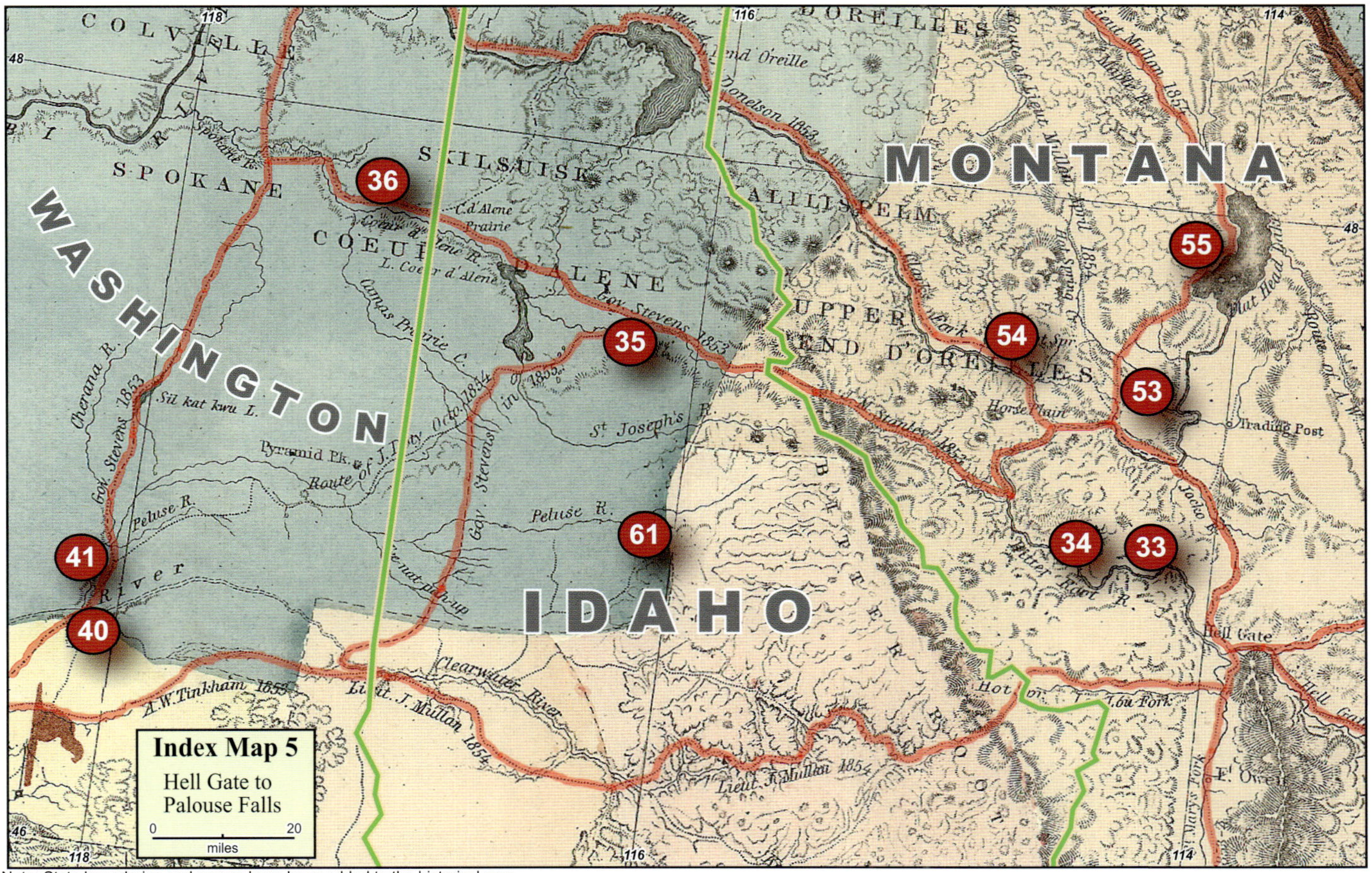

Note: State boundaries and names have been added to the historical map.

In some places, small rivers and streams follow the channels, such as the Palouse River. A spectacular waterfall marks the flow of the Palouse just before its juncture with the Snake River. Here the water drops nearly 198 feet vertically into a narrow deep canyon lined by basaltic rock.

The weather began to deteriorate. It was midfall; temperatures were dropping, and chilling winds were picking up. The long expedition had taken a toll on Stevens's health (aggravating wounds he received during the Mexican War), and he was anxious to meet up with McClellan and the western survey party. He made this connection at Fort Colville, a Hudson's Bay Company trading post east of the Columbia River, just north of Kettle Falls (see index map 6).

Eventually, the main party reached another trading post, Old Fort Walla Walla, on the Columbia's eastern bank at its junction with Walla Walla River. From there the men traveled by small boat and canoe to Fort Dalles, then downstream through the vast Columbia River Gorge. The gorge was created by the river cutting across the Cascade Mountains as the system gradually uplifted. They passed the site of a huge landslide that forced the river's course southward into a narrow channel. It was here they encountered dangerous rapids. Today the Bonneville Dam subdues this once hazardous portion of the river.

After passing Cape Horn, Stevens reached Fort Vancouver in November 1853. The fort was a large complex consisting of a trading post operated by the Hudson's Bay Company and a military post known as Columbia Barracks. Stevens's journey was nearly over.

The main party moved north to Olympia, Washington, where Governor Isaac Stevens began his new job as head of Washington Territory. Not one to sit still for long, within a month Stevens explored Puget Sound, investigating its hydrography. Later in 1855, he conducted an important treaty-making effort with major tribes around Puget Sound and in Idaho and Montana. These treaty negotiations became defining and controversial events in the region's history during the next decade.

Scablands and Palouse Hills —Map by Philip Mobley

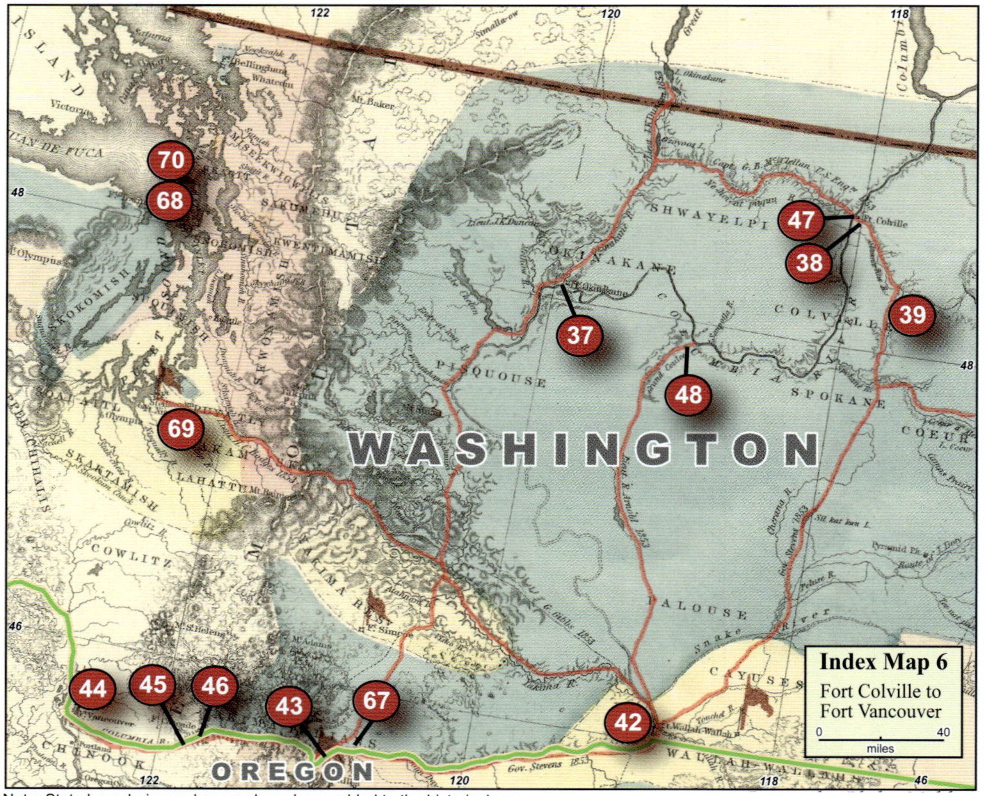

Note: State boundaries and names have been added to the historical map.

Part 3

Gallery of Views

St. Paul, the rendezvous point for Isaac Stevens's survey party, is the report's first illustration. Stevens arrived in St. Paul by steamboat on May 27, 1853, after a five-day journey up the Mississippi River from St. Louis.

St. Paul was six miles from Fort Snelling, a major military post constructed in the early 1820s at the confluence of the Minnesota and Mississippi Rivers, just downstream from St. Anthony Falls (the present site of Minneapolis). This junction was the effective head of navigation for steamboat traffic on the Mississippi River, making it a logical departure point for westward exploration and travel.

Originally settled in 1838–40 as a trading post named "Pigs Eye," St. Paul grew in importance and was designated Minnesota's territorial capital in 1849. Although the city was still in its infancy, Stevens was impressed:

> St. Paul (see accompanying sketch) is beautifully located upon a high bluff on the east bank of the river, and is rapidly growing in size and importance. It is quite a business place, everything indicating vigor and activity. Among its prominent buildings are the territorial capitol, modelled after the Capitol at Washington, and several very fine churches.[13]

The scene depicts the Mississippi River flowing placidly by the sparsely settled town. Yet the beginnings of economic prowess are suggested by the inclusion of a steamboat and a flatboat. The detailed depiction of the flatboat is another interesting component of this image. Such large craft were difficult to steer. The drawing depicts five "sweeps," frequently referred to as "broadhorns," which were used to control the boat's direction.

Interior view of Fort Snelling today, St. Paul, Minnesota
—Courtesy of Kevin Hawkins and used by permission

Plate I. *Saint Paul.* Artist: Stanley.

Minnehaha Falls —Courtesy of Kevin Hawkins and used by permission

WHILE ORGANIZING HIS SURVEY PARTY in the St. Paul area, Stevens set up a base camp on Lake Amelia, three miles northwest of Fort Snelling. He named his temporary headquarters Camp Pierce in honor of newly elected president Franklin Pierce.[14] Near the camp, the survey party encountered a picturesque waterfall, which Stevens described in detail:

> One of the curiosities of our vicinity, which was sketched by Mr. Stanley, and which is shown in the accompanying sketch, (No. 2.) is the Minne-ha-ha or the Laughing Water, called also Brown's Falls. . . . Ten miles above the Falls the stream flows from Lake Calhoun, and it passes through a level but fertile prairie, its margin decked with a wholesome growth of willow, poplar, and hazel, while at a short distance there are little forests of black-jack (*Quercus imbricaria, Michaux*) and other trees of like character. Here the sheet of water is from twenty to twenty-five feet wide, and its fall forty-one feet. The rock over which it pours shelters an oval cave about seventy-five feet wide and thirty feet from the falling water to the back. The view is taken from beneath this shelving rock. Though the magnitude of this cascade is not such as to excite our wonder, its picturesque beauty and pleasing melody attract the admiration of every visitor. . . . The soil above this valley is very rich, and there are good grounds for the opinion that the axe, the plough, and the mill-wheel will before long invade and materially change the character of the scene.[15]

Stanley's depiction of the falls reflects peace—a sense of Eden. In contrast, as Stevens predicted in his report, today the site is in the midst of urban Minneapolis, in a residential neighborhood dotted with single-family homes and lined by parkways and gridded streets. In the image's foreground are two men, possibly Stevens and Stanley, enjoying the falls by fishing and painting, respectively. Because the men are noticeably smaller than the waterfall, which has a drop of forty feet, their inclusion helps the viewer appreciate scale.

The place-name, "Minnehaha," which was cited in Henry Wadsworth Longfellow's 1855 poem *Song of Hiawatha,* was the product of combining two Native American (Dakota) words: *Ha ha,* identifying a waterfall, and *Minne,* meaning water. The other, earlier name for the falls, Brown's, referred to the commander-in-chief of the U.S. Army in 1823, Jacob Brown.[16]

Plate II. *Minnehaha or Brown's Falls near Fort Snelling.* Artist: Stanley.

LEAVING THE ST. PAUL–FORT SNELLING AREA on June 6, the survey party traveled north approximately seventy miles along the Mississippi River to Sauk Rapids and the Sauk River. They established another temporary camp on the north side of the Sauk River, about two miles upstream from its junction with the Mississippi. They named it Camp Davis in honor of Secretary of War Jefferson Davis.

The party arrived at this location in three detachments, two traveling by land on either side of the Mississippi, and the third by steamboat, which the men had engaged above St. Anthony Falls. They encamped at Camp Davis for approximately ten days while waiting for further supplies. Advance parties explored routes westward along the Sauk River.

Stevens left Camp Davis on June 16, traveling eighteen miles west along the Sauk River to a spot where the advance party had determined the river could be crossed. Interestingly, India rubber boats were used to ford the Sauk and a few other streams. Stevens recounted the crossing:

> Sauk river (see accompanying sketch) at our ford is about 120 feet wide, though, owing to the obliquity of the banks and rapidity of current, the ford is near 300 feet wide, and the water five feet deep. . . . In the crossing of the Sauk by the main train the India rubber boats were, for the first time, used. The larger one is about twelve feet long and four wide, weighing seventy-five pounds, the other about one-fifth smaller. A rope was stretched across the stream and the boats ferried across by means of a ring attached to their bows and sliding along the rope. They succeeded admirably; and a birch canoe, managed by one of the voyageurs, was also used in crossing. Some of the men were in the water for hours, but worked faithfully and efficiently.[17]

The Sauk, a major tributary river of the Mississippi, begins at Osakis Lake and flows southeasterly 120 miles, joining the Mississippi near Sauk Rapids. It provides an easy route through central Minnesota heading northwest toward the Red River of the North.

Plate III. *Sauk River.* Artist: Stanley.

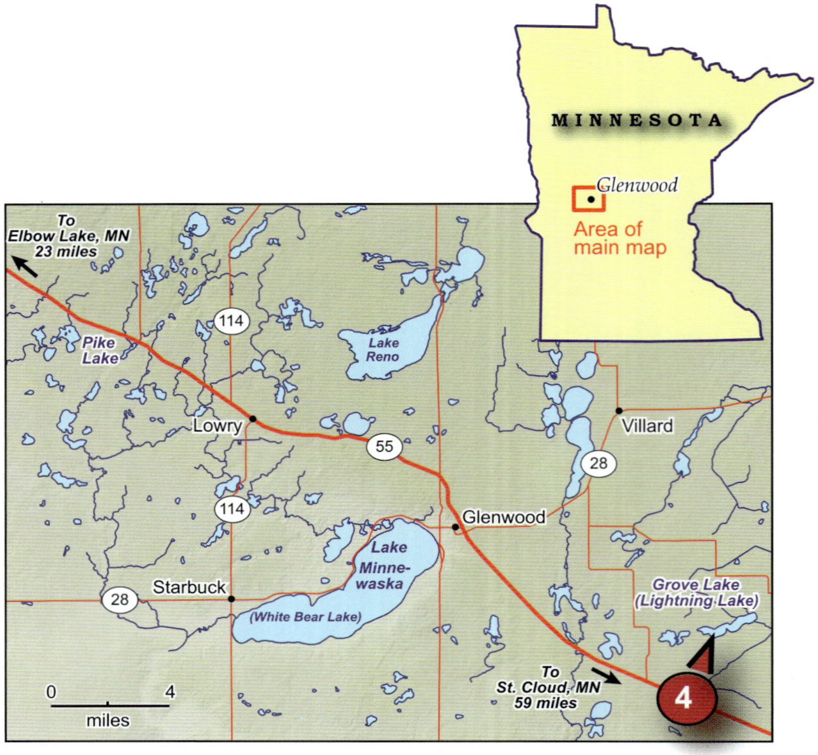

EXPLORATIONS SELDOM GO INTO completely unknown territory. As Stevens's party moved northwest, they were using information acquired by a U.S. Army Corps of Topographical Engineers expedition led by John Pope and Samuel Woods in 1849. They were also using part of a thoroughfare called the Red River Trail. This trail connected settlements along the Red River with trading posts and businesses at St. Paul and Fort Snelling.

When the Pope-Woods party camped on the shores of Lightning Lake, they were buffeted by a violent storm. Lightning ravaged the group, and one man was killed. The lake was named for this event. Today known as Grove Lake, it is about one mile west of the town of Grove Lake in Pope County, Minnesota.[18]

Stevens's party camped on its shores on June 19, which was a Sunday. The group took time to rest and to repair equipment and clothing. Stevens described the evening:

> Arrived at camp on Lightning lake about 8 p.m. . . . Distance travelled, eighteen and three-quarter miles. The frequent sloughs and bad crossings in our march to-day added much to the labors of the men and animals. . . . After the hard day's march we enjoyed our supper of game, cooked in hunter's style on sticks before the fire, although it was midnight before we could have it ready. . . . Lightning lake (see accompanying sketch) is a very beautiful sheet of water. . . . Its northern shore is thickly studded with timber, and the southern side, upon which we are encamped is more thinly wooded, but affords an ample supply for all camping purposes. The water is excellent, as is also the grass. Pickerel, pike, and bass fill the lake, numbers of which our parties caught; and ducks, geese, swan, plover, and prairie chicken, abound in the vicinity.[19]

Today, Minnesota is well known for its ten thousand lakes. Not surprisingly, several of these lakes were featured on the lithographic prints reproduced in the report's first pages. Most of them, known as "kettle lakes," were the product of continental glaciation. They formed when huge ice masses swept over the northern part of North America, removing and depositing rock, large erratics, and smaller debris on the landscape. As the last major glacier retreated, most likely in irregular moves, it lost large blocks of ice, the size of which are difficult to imagine. When the ice melted, a depression was created in which water then collected.

Plate IV. *Lightning Lake*. Artist: Stanley.

AFTER LEAVING LIGHTNING LAKE, the party's next stop was White Bear Lake, now known as Lake Minnewaska, located just west of present-day Glenwood. Members of the Pope-Woods expedition recorded this lake as White Bear in 1849. White Bear, or Waube-Mokwa, was an Objibway chief who lived in the region several hundred years before.[20] While camping by its shores, Stevens took note of the landscape around this attractive lake:

> We arrived at White Bear lake, about nine and a half miles from this morning's camp, at 10 a.m. Leaving Lightning lake the country seems to change its character and is no longer a flat, undiversified surface, with occasionally a gentle undulation scarcely attracting attention. It has gradually changed to a heavy rolling prairie which, before approaching White Bear lake, becomes broken up into hills, valleys, and basins, varying from thirty to fifty feet in depth. Boulders and stones, from the size of pebbles to paving-stones, are very numerous. . . . White Bear lake, (see accompanying sketch,) upon or near which most of the parties of the survey are encamped, lies in sight of our trail, about two miles distant to the south. It is a beautiful sheet of water, bordered with timber, about fourteen miles long and two wide, with high, swelling banks running back a mile or so, and rising to the height of about one hundred and fifty feet.[21]

Stanley depicted a placid scene near the lake. Two wagons were shown, along with a cart transporting a canoe in which a man sat. This type of cart, known as a Red River cart or Pembina cart, was typically used in regions where the soil was soft and traction was difficult. To counter the problem, the cart was constructed with wide, hide-covered wheels.

In this chromolithograph, Stanley used a strong red-orange in the foreground and neutral grays in the background. He positioned the men and horses in the image's lower right, utilizing a power point, and placed them on the large hill dominating the foreground. The result is an image with good depth, or three-dimensionality.

Red River, or Pembina, cart —Drawing by Philip Mobley

Plate V. *White Bear Lake.* Artist: Stanley.

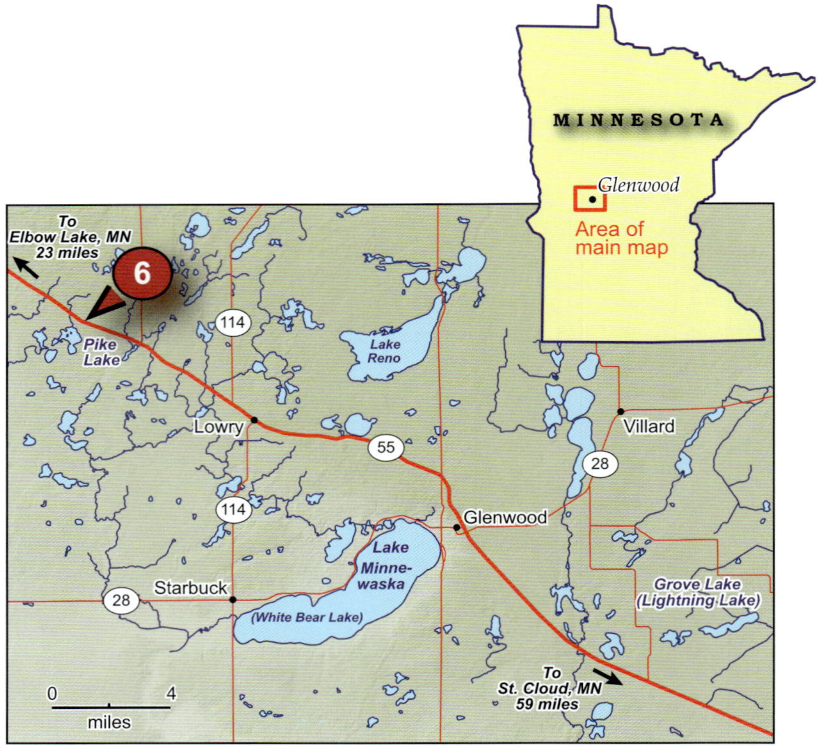

DURING LATE JUNE OF **1853**, the party continued its northwestward plod through central Minnesota as the weather turned abnormally wet. Stevens observed how the rainy weather affected their travels:

> The season was unusually backward, the rains frequent and heavy, and great labor was required in crossing the swollen streams—some by bridging, others by means of the india-rubber floats for ferrying over the goods while the animals swam across. The wagons, bogged in the miry ground, had to be frequently unloaded and loaded again; but many soft and marshy places were made passable by covering the road deep with cut grass.[22]

The third lake illustrated in the report was Pike Lake. Such lakes were good locations for encampments, providing both fresh drinking water and fishing. This lake, in the Nora township of Pope County, was named by the Pope-Woods expedition in 1849 for the many pikes the men caught there.[23]

Stevens arrived at Pike Lake on June 23, where Lt. Cuvier Grover, in charge of an advance party, had already established a camp:

> Lieutenant Grover's camp is beautifully situated on the shores of the lake. The main train . . . arrived about half-past eight, and the whole expedition was again brought together. This I consider as the real starting point of the expedition, and named our camp here Camp Marcy, in honor of the Secretary of State. We remained here a day, in order to give the animals a chance to rest. They appear to be in very good condition, and the grazing is fine. Received of the various scientific chiefs reports of their labors to this point.[24]

The scene depicted here is a quiet one. The camp was split into two or more sections. With parties of this size, camps were often separated for sanitary reasons. It was common to divide camps according to military rank—officers in one camp, enlisted men in another. In this illustration, one small group of tents was positioned at the lake's edge, while another was placed along the lakeshore in the background. Everything in the image reflects calm. But was this serenity a true reflection of expedition life? Quite possibly not. At this point in the trip, the men were told that they might not have sufficient rations to reach Fort Union. Not surprisingly, the men grumbled and threatened to "strike":

> On counting rations, it was found that for the main party there was a supply for twenty days, while it might take forty-five to reach Fort Union. But with the eight oxen in the carts, and the known abundance of game, I feared no scarcity. The men showed some anxiety and talked of a strike, but, seeing the confidence of the officers, abandoned any open demonstration. Previous to this time I had ordered a reducing of rations whenever the quantity of game would justify it, and henceforward I gave the most particular attention to it, so that, although we did not reach the Yellowstone for thirty-eight days (August 1), there was at no time a scarcity of provisions.[25]

Plate VI. *Pike Lake.* Artist: Stanley.

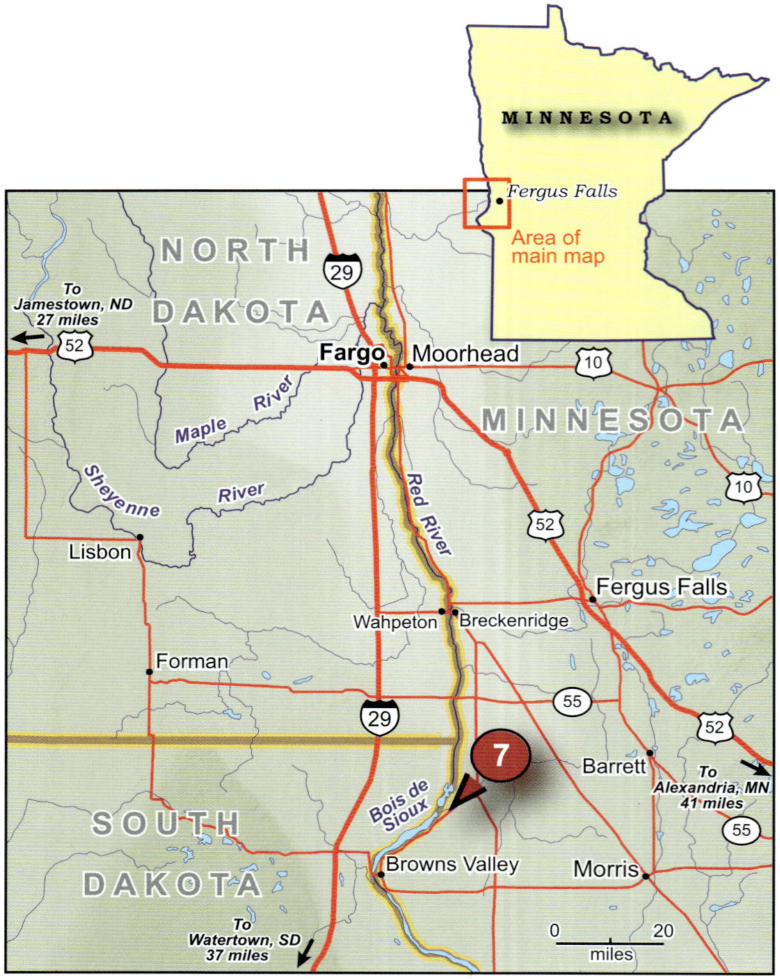

FROM PIKE LAKE, THE SURVEY PARTY CONTINUED in a northwesterly direction, still following the Red River Trail. The men struggled with soft, wet ground miring the wagons until reaching the Bois de Sioux River on June 28. Upon arriving at this benchmark, the men, including Stevens, were relieved:

> Numerous large catfish were caught this afternoon, some weighing from 12 to 20 pounds. At 11 p.m. we sat down to a supper of ducks, catfish, and coffee, and all the men were in fine spirits. The Bois de Sioux had been a great point to reach—the end of bad roads and the commencement of the buffalo country. It is quite a large stream, is well lined with wood, and the drift deposits on its banks indicate that during spring freshets a considerable portion of the prairie is submerged.[26]

This is another very placid landscape. Trees and smaller scrubs provide eye-catching appeal. The river appears to be gently flowing, but considering the wagons' position, getting down the steep banks would be a chore. The sky's depiction is dramatic—a good example of luminescence.

The Bois de Sioux is a generally northward-moving stream. Flowing out of Lake Traverse, it runs north-northeast until it joins the Ottertail River to form the Red River of the North. Today the Bois de Sioux and the Red River define the border of Minnesota with South Dakota and North Dakota.

Plate VII. *Bois de Sioux River*. Artist: Stanley.

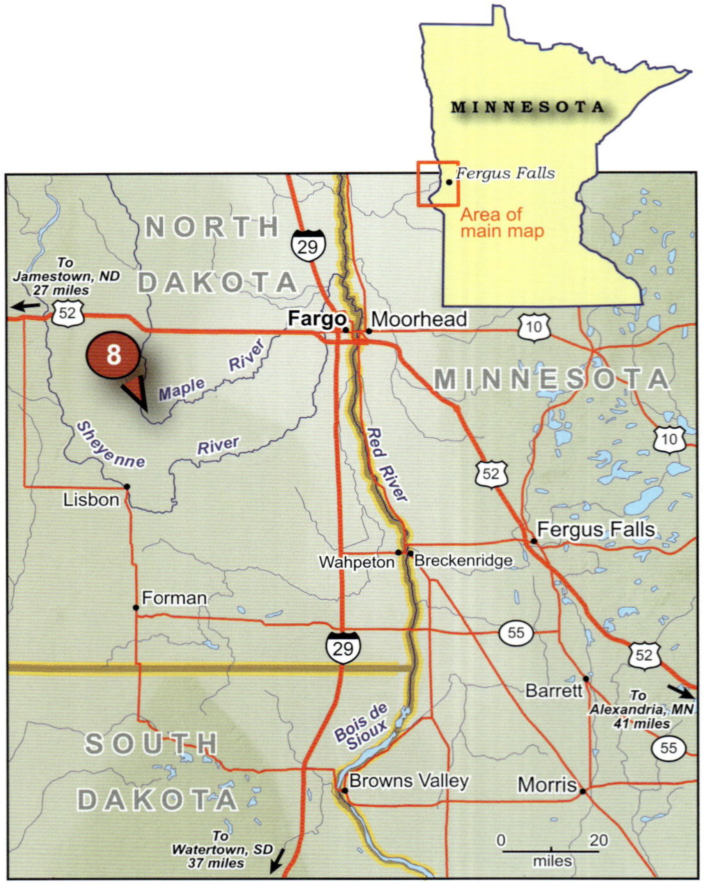

Plate 29 of Stevens's zoological report, illustrating a number of different frogs collected on the expedition —Courtesy Geography and Map Division, Library of Congress

After crossing the Bois de Sioux, the party continued on a northwestward route through today's eastern North Dakota, where they crossed numerous tributaries of the Red River. One of these was the Sheyenne River, which they crossed twice. The party first crossed the Sheyenne on July 2 and again six days later (see plate 9). Another was the Maple River, which the party arrived at and crossed on July 4, after traveling most of the day. Stevens briefly described the trek:

> Got off about 10 and followed the Red River trail some twelve miles, when we left it altogether. Crossed Maple river and camped on its banks, about half a mile from our crossing. (see accompanying sketch.)[27]

To celebrate the nation's birthday, the men contributed red and white shirts to create a makeshift flag, sewn together by Henry Boulieau, one of the French voyageurs who joined the expedition after the party left St. Paul. At dusk, the men gathered around the flag and gave three hearty cheers. Later, they "indulged in some refreshments in honor of the day."[28] One assumes that they drank a medicinal tonic reallocated for this purpose. Storytelling and singing continued through the evening hours.

Until the men reached Maple River, they were following the Red River Trail, which traversed known territory. As the country became less familiar, the explorers paid more attention to collecting scientific information. They calculated the longitude and latitude of different points and took serious notes on the landscape, flora, and fauna. Plants and animals were observed, and samples were collected for documentation and further analysis.

Stevens noted that frogs were plentiful in the region, which possessed an abundant supply of water and food in the form of insects. Today, however, populations of frogs and other small amphibians are rapidly decreasing. These fragile and sensitive critters are probably dying from various causes, including acid rain, marshland elimination, water pollution, and global warming.[29]

Plate VIII. *Maple River.* Artist: Stanley.

On July 8, the party made their second crossing of the Sheyenne River, where this landscape view was sketched. Stevens recorded the scene in his narrative:

> Started this morning at 6 and arrived at the crossing of the Shyenne river after a march of about fifteen miles. . . . We went into camp about one o'clock, on the east and south side of the Shyenne, (see accompanying sketch,) and a party was at once detailed to cut wood and prepare charcoal. . . . As the grass here is very indifferent, the main train will cross the river to-morrow afternoon and go ahead some six miles to good grass. . . . Travelling along the côteau of the Shyenne, near the place where we encamped, I saw a buffalo, to which Menoc and La Frambois gave chase, but without success. The plateau of the river, and the coulée opening into it, is observable to-day. . . . The grass, at the best, is very poor, and the great abundance of grasshoppers had made sad havoc with what had grown here.[30]

The Sheyenne River, like several other streams in this region, forms an unusual drainage pattern. As it flows 325 miles in a southeasterly direction through eastern North Dakota, it follows a rough U-shaped course. Toward its end, it bends northward as it joins the Red River, which also flows in a northerly direction. This unusual drainage pattern is a by-product of continental glaciation. When the glaciers retreated and the ice wasted away, a depression remained in the crust. Since water flows downhill, the rivers moved toward this depression. For a while this basin was filled by a huge glacial lake known as Lake Agassiz. Today's Lake Winnipeg in Manitoba is a remnant of the ancient lake.[31]

Even though the river crossings that Stanley depicted in Plates 7 through 9 are located some distance apart, they provide a rough definition of the route the party followed through what is now southeastern North Dakota. These views illustrate campsites the party used during an eleven-day period.

The Sheyenne River in North Dakota
—Photograph by Jannell Best Rundell and used by permission

Plate IX. *Shyenne River.* Artist: Stanley.

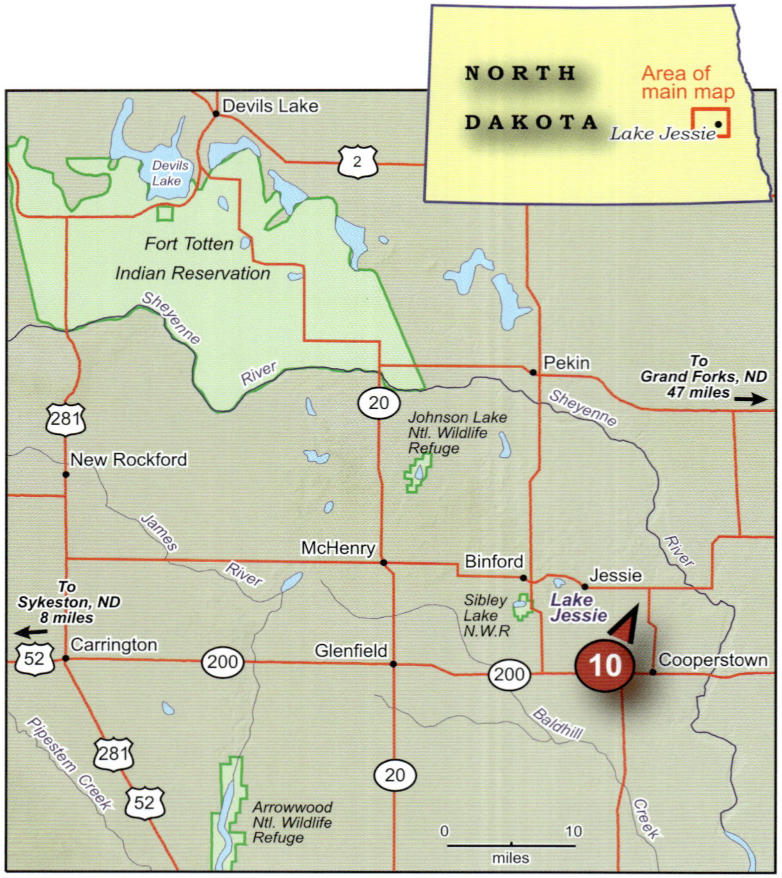

On July 10 the survey party encountered their first large herd of bison near Lake Jessie. They were astonished by its immense size:

> About five miles from camp we ascended to the top of a high hill, and for a great distance ahead every square mile seemed to have a herd of buffalo upon it. Their number was variously estimated by the members of the party—some as high as half a million. I do not think it is any exaggeration to set it down at 200,000. I had heard of the myriads of these animals inhabiting these plains, but I could not realize the truth of these accounts till to-day, when they surpassed anything I could have imagined from the accounts which I had received. The reader will form a better idea of this scene from the accompanying sketch, taken by Mr. Stanley on the ground, than from any description.[32]

Stanley's depiction of this herd is one of the most reproduced views from the Northern Railroad Survey report. It depicts thousands of these woolly grazers overwhelming the landscape. Since they were being viewed from afar, it is hard to appreciate their true size.

As the survey team proceeded west, the men encountered many more buffalo. Members of the party frequently hunted the animals, making bison meat the usual main course around the evening campfire.

The countryside east and west of Fort Union was filled with bison. Consequently, the region was a great slaughtering field for these magnificent creatures. The hides were a major trade good at the fort. In peak periods, it was not unusual for traders to ship 100,000 or more skins down the Missouri from the post.[33]

Herd of bison —Photograph by Paul McDermott

Plate X. *Herd of Bison near Lake Jessie.* Artist: Stanley.

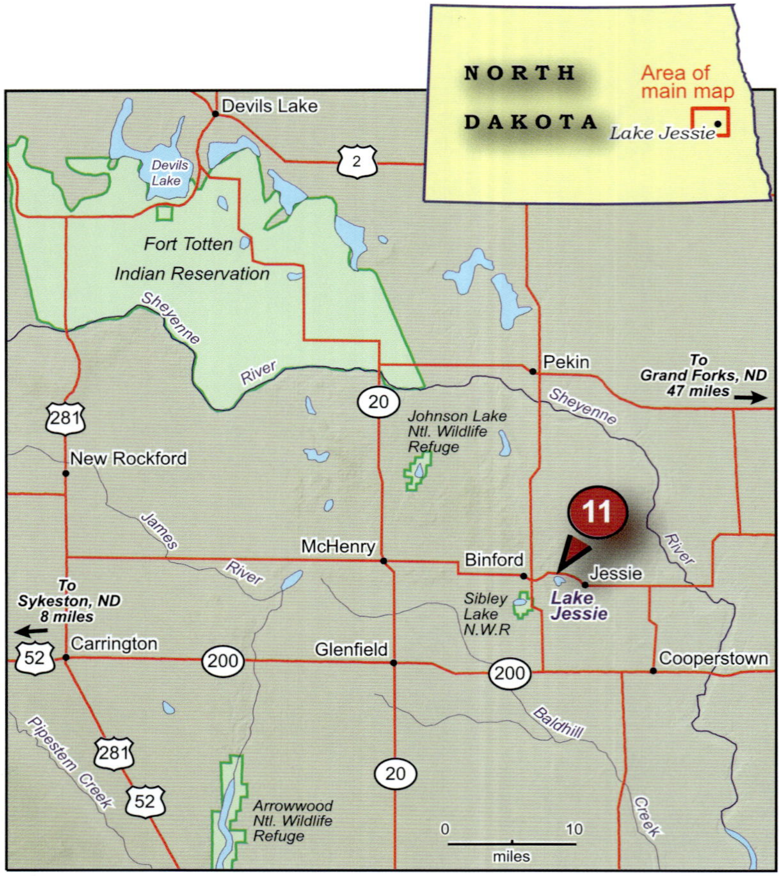

As the Stevens party moved into the high plains, potable drinking water for men and animals became a scarce commodity. In the vicinity of Lake Jessie, where the buffalo herd was sighted, numerous lakes punctuated the landscape, but the water was seldom potable. Lake Jessie, Stevens noted, possessed poor water quality. Yet the party soon found relief in the discovery of a spring:

> We arrived at Lake Jessie about 3 o'clock p.m., the bluff shore on which we encamped being some sixty-four feet above the level of the lake (see accompanying sketch.) . . . The water of Lake Jessie is considerably saline in its character, but about three-fourths of a mile from camp an excellent spring of good fresh water was found by Henry Boulieau and myself while out on a reconnoitring trip. This discovery afforded a pleasant relief from the brackish water of Lake Jessie and the surrounding lakes.[34]

Lake Jessie is located between the Sheyenne and James Rivers in central North Dakota. The present town of Lake Jessie is much farther west, about twelve miles southeast of Williston, but the lake Stevens visited is situated among several lakes in the vicinity of today's Jessie, North Dakota.

Besides drinking quality, water in this region posed another problem. Ponds, lakes, and marshes created breeding grounds for noisome insects like mosquitoes and flies. Throughout their journey westward, expedition members were horribly tormented by these pests:

> The mosquitos are exceedingly annoying, flying against the sides of the tents with a noise like the pattering of rain, while the inside is perfectly black with them. Their constant humming drove the men out into the open air, and rendered it almost impossible to sleep.[35]

Plate XI. *Lake Jessie.* Artist: Stanley.

From Lake Jessie, Stevens's party followed a westward course that roughly paralleled the Sheyenne River. On July 16, having progressed approximately sixty to seventy miles beyond Lake Jessie, they encountered a group of Métis known as the Red River Hunters. Their train consisted of 1,300 people, 824 carts, and 1,200 animals. Camping nearby, Stevens described the Hunters' encampment:

> They encamped near by, and the close yard which they formed presented quite a contrast to the open camp adopted by us. They made a circular or square yard of the carts, placed side by side with the hubs adjoining, presenting a barrier impassable either to man or beast. The tents or lodges were arranged within, at a distance of about twenty feet from the carts, and were of a conical shape, built of poles covered with skins, with an opening at the top for the passage of smoke and for ventilation. They were 104 in number, being occupied generally by two families, averaging about ten persons to the lodge. Skins were spread over the tops of the carts, and underneath many of the train found comfortable lodging places (see accompanying sketch).[36]

Stanley captured the richness of the Hunters' culture in this detailed scene, and in a relatively lengthy discourse covering nearly three pages, Stevens described this fascinating people.[37] The Red River Hunters were on their first hunt of the season, looking for buffalo. In the fall, they would again venture onto the plains seeking their quarry, especially to obtain good robes. The assemblage began its journey near Pembina Mountain in Canada and traveled south through what today is known as Devils Lake, North Dakota. At Pembina, the group practiced agriculture, growing such crops as potatoes, wheat, corn, and barley.

The Red River Hunters were characterized by diversity. Most were of mixed blood; the rest were either full-blood Native American or French-Canadian. Their colorful dress made an impression on Stevens:

> The men dress usually in woollens of various colors. The coat generally worn, called the Hudson Bay coat, has a capot attached to it. The belts are finely knit, of differently colored wool or worsted yarn, and are worn after the manner of sashes. Their powder horn and shot bag, attached to bands finely embroidered with beads or worked with porcupine quills, are worn across each shoulder, making an X before and behind. . . . The women are industrious, dress in gaudy calicoes, are fond of beads and finery, and are remarkably apt at making bead work, moccasins, sewing, etc.[38]

Plate XII. *Camp Red River Hunters.* Artist: Stanley.

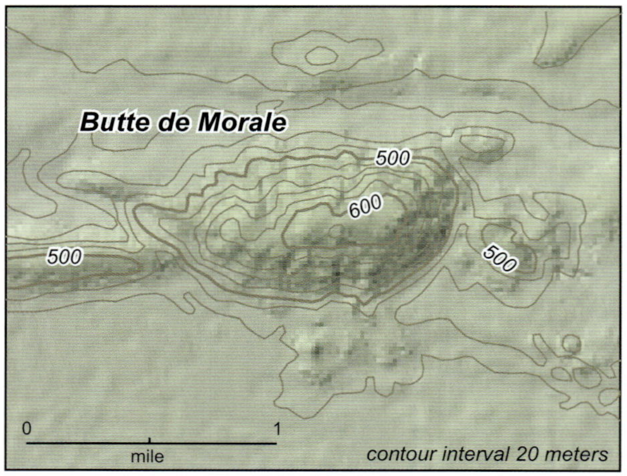

Topographic map of Butte de Morale

THE FOLLOWING DAY, AFTER CAMPING WITH the Red River Hunters, the survey party passed Butte de Morale, a prominent landform just northwest of the Sheyenne River in central North Dakota, between the present towns of Selz and Harvey. It lies approximately at latitude 47° north and longitude 99° west. Rising to a height of 1,775 feet, its crest provides a beautiful view. Certainly during the buffalo-hunting period, many Native Americans and fur trappers scaled its slopes to take advantage of the elevated perspective.

Stanley, too, hiked the slope. His image, rendered from one of the higher vantage points, reveals an idyllic scene. The foreground is filled with crooked vegetation, while Goose Lake and a hill are faintly visible in the background. Its composition and rich coloration make this view one of the three most picturesque illustrations in the collection.

Look closely at the background, especially in the watercolor. Surrounding the lake is small herd of buffalo, which appear as small dots adjacent to the lake. Tucked away on the image's left side, in a small pocket, are three army tents—part of Stevens's camp.

This small butte was named after Morale, a French-Chippewa hunter killed by the Sioux Indians about 1840.[39] In Stevens's time, the hill marked the core of the buffalo country exploited by the Red River Hunters and others. The hunting activity left horrific remnants scattered across the grassland:

> One of our men reported having rode through a section of land, some quarter of a mile square, on which were strewn the remains of some three hundred buffaloes. In killing these animals, the choice pieces and hides are only taken, while the remainder is left as a prey to wolves or to rot on the prairie.[40]

Stevens had the butte's altitude measured on July 18, finding that its relative height above the surrounding land was 281.8 feet. Lake Stevens, which is seen in the image, derives its name from the governor. The lake is also referred to as Goose Lake because of its narrow, elongated form, which stretches four and a half miles.

Butte de Morale in early spring —Photograph by Jannell Best Rundell and used by permission

Plate XIII. *Butte de Morale.* Artist: Stanley.

Manuscript watercolor of Butte de Morale by John Mix Stanley —Courtesy of Yale University Art Gallery and used by permission

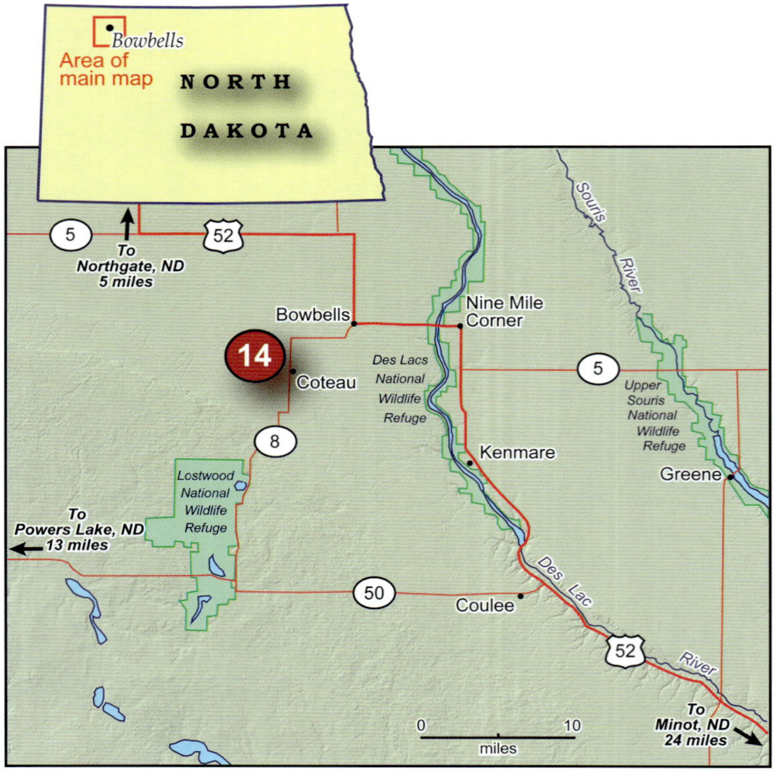

ON JULY 27, WHEN THE SURVEY PARTY was about 100 miles away from Fort Union, they encountered a camp of Assiniboines (pronounced uh-SIN-uh-boyns). After dinner Stevens, accompanied by his interpreters, went to their camp. He marked this visit with a detailed description:

> [The camp] was irregularly arranged in a sort of corral, consisting of about 150 lodges, averaging ten persons to each lodge. Our approach was hailed by the barking of an immense number of dogs. These dogs are a prominent feature in every Indian camp, being used for drawing lodges, provisions, and property, from place to place, indeed furnishing the entire transportation of the Indians in winter.[41]

The Assiniboines are a northern Plains tribal group. In the mid-nineteenth century, they lived along the Upper Missouri and ranged into Saskatchewan, Canada. Their language was Siouan, but they had separated from other Sioux peoples in the mid-1600s.[42] When Stevens visited the Assiniboines in 1853, they identified themselves as buffalo hunters. During the year, they migrated between campsites, hunting animals for food and skins.

Stevens's documentation of the Assiniboines was sensitive and insightful. He portrayed a people that were well aware of the changes taking place in their environment. Even though buffalo were still plentiful in the 1850s, the tribe was aware that the animals were rapidly disappearing, and their leaders recognized that they would soon be facing economic hardship:

> My father, you see us now as we are. We are poor, we have but few blankets and little clothing. The Great Father of Life who made us and gave us these lands to live upon, made the buffalo and other game to afford us subsistence; their meat is our only food; with their skins we clothe ourselves and build our houses. . . . But I fear we will soon be deprived of these; starvation and cold will destroy us. The buffalo are fast disappearing and before many years will be destroyed. As the white man advances, our means of life grow less. . . . My father we hear that a great road is to be made through our country. We do not know what this is for; we do not understand it; but I think it will drive away the buffalo.[43]

Plate XIV. *Distribution of Goods to Assiniboines.* Artist: Stanley.

SEVERAL DAYS BEFORE MEETING WITH THE ASSINIBOINES, the survey party approached the Mouse River. This river, also known as the Souris River, parallels the Missouri in this part of North Dakota. It begins in Saskatchewan and stretches 450 miles in a broad loop before entering the Assiniboine River, which flows back into Canada.

The Mouse River valley in 1853 was textured with trees such as oak, ash, and maple. The valley was etched into the landscape, lying at least 200 feet below the surrounding prairie. Here the trees were protected from wind damage. Stevens referred to such valleys as *coulées*, from the French word for "flow" referring to valleys, which in the northwestern United States were generally created by erosion from glacial meltwater.[44] The best-defined coulees the expedition encountered were farther west, on the Columbia River Plateau (see plate 48, "Grand Coulee").

Stevens took note of the valley's peaceful beauty:

> The configuration of Mouse River valley, as well as its tributaries, resembles that of the Shyenne (see sketch.) High ridges divide the plateau bordering the stream from that extending into the prairie, with coulées intersecting it and opening into the river on one side, gradually growing imperceptible as they make into the prairie on the other. . . . Many of the coulées reach to the edge of the Missouri plateau; and in the examination for a good passage for the wagon train, secluded spots were found where beetling crag and winding stream, venerable trees and greenest sward combined in scenes of much picturesque beauty.[45]

Stanley's depiction of the scene captures much of the romance of Stevens's commentary. Trees are tucked into the nooks and crannies along the valley. Grassland abounds. Stretching from the flat plateau in the background, a long, sinuous wagon train moves down into the valley. The train was viewed looking toward the southeast. The exact location of the artist's position is not known, but it was probably near present-day Sawyer, North Dakota. The soothing application of lines defining hill shapes and slopes evokes a gently rolling sea. The serenity is interrupted only in midground by the vertical sedimentary rocks protruding from the river's edge.

Plate XV. *Near Mouse River.* Artist: Stanley.

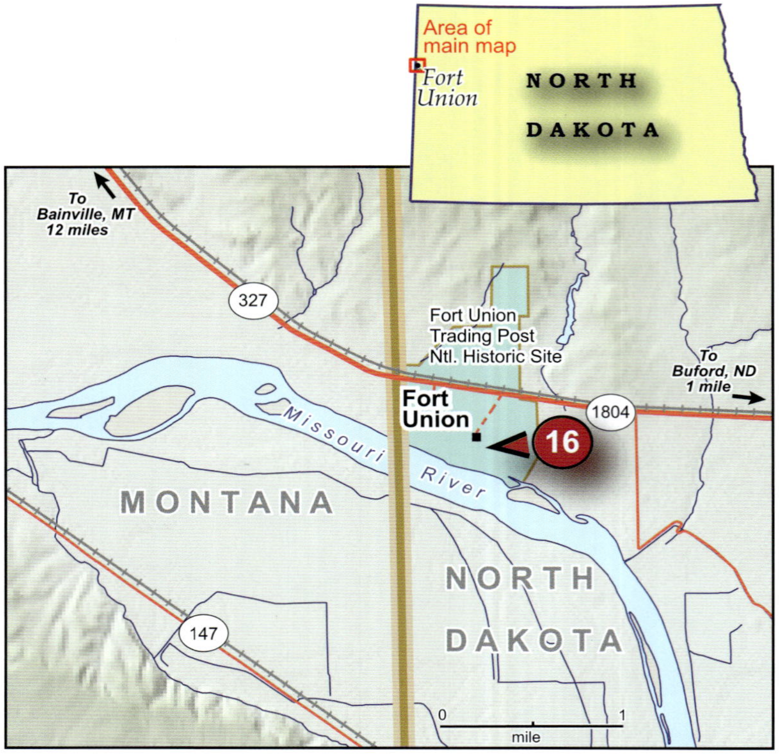

On August 1 Stevens arrived at Fort Union, where the party remained nine days, resting and replenishing supplies. It had taken fifty-five days to reach the trading post from St. Paul. The total distance traveled up to this point was 715 miles, with the party averaging about 15 miles per day.

The fort was strategically located a short distance upstream from the junction of the Yellowstone and the Missouri Rivers, not far from today's Montana–North Dakota boundary. Major river junctions were attractive locations for establishing trading posts throughout the West. Many eventually evolved into large settlements. The American Fur Company post quickly became a major trading site; in fact, some people refer to Fort Union as the "Times Square of the Great Plains."[46] However, this particular post did not grow into a city or even a small settlement. Today it is the site of the Fort Union National Historic Site, run by the National Park Service.

When Stevens's party arrived at the fort, trading activities were at their height.[47] The governor gave a detailed description of the fort:

> Fort Union (see sketch) is situated on the eastern bank of the Missouri river, about 2 miles above the mouth of Yellowstone. It was built by the American Fur Company in 1830, and has from that time been the principal supply store or depot of that company. It is framed of pickets of hewn tim-

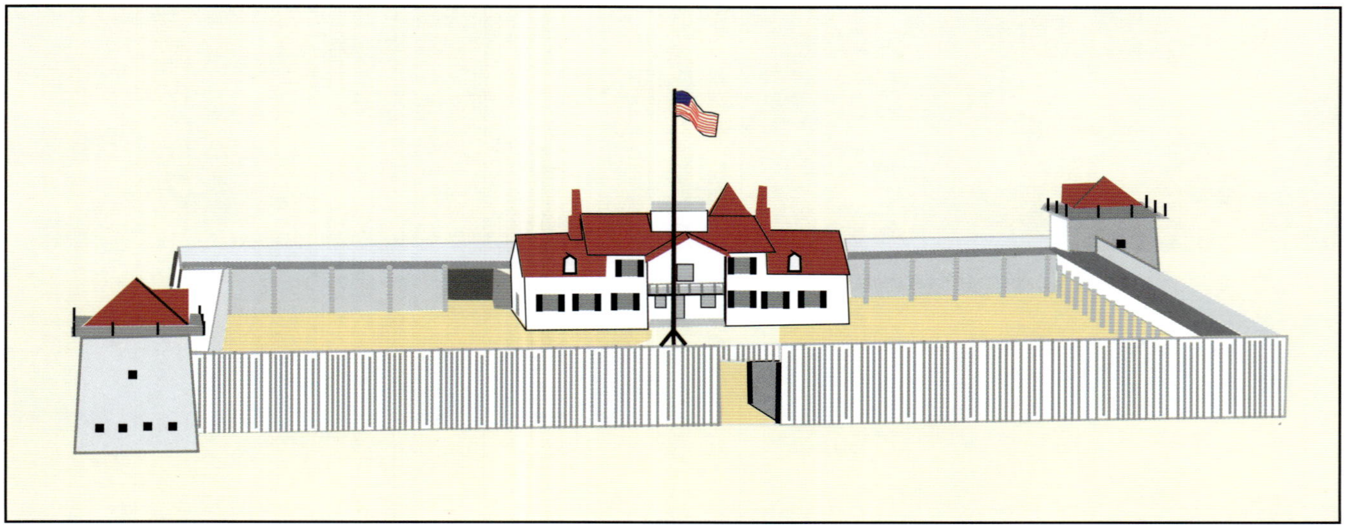

Plan of reconstructed Fort Union —Drawing by Philip Mobley

ber, about 16 feet high, and has two bastions, one at the northwest and one at the southeast corner. The front or main entrance is on the side opposite the river. This fort is probably 250 feet square. The main buildings, comprising the residence of the superintendent and the store, are on the front or eastern side. They are two stories high, and built of wood. The shops and dwellings of the blacksmith, the gunsmith, the carpenter, the shoemaker, the tailor, and others are of adobe or of wood, and occupy the other sides. These mechanics are mostly French half-breeds.[48]

Stanley's portrayal of Fort Union, drawn from a vantage point northwest of the fort, is fascinating. One cannot see much of the post's interior, but it is apparent that the bastions contain wind vanes, and one building's roof has a platform upon which a flagpole was positioned. In the foreground, outside the fort, Stanley documented their second distribution of gifts to the Assiniboines (see also plate 14). Two groups are seated beneath distinctive structures—skin- or blanket-covered shelters that form a shield from the wind and sun, supported at two ends by tilted poles. Trade goods are scattered around several tepees, and a few items are positioned near men conversing in the foreground.

A plan of the fort is illustrated here. The most impressive component was the large mansion inside. This was the home and office of the factor, referred to as the Bourgeois House.

In 1867 the fort was completely dismantled. In the late twentieth century, after considerable research and archeological excavation, the fort's original character was determined, and the trading post was reconstructed over a four-year period ending in 1989. It now stands as a faithful replica of the original post. The rebuilt Bourgeois House serves as museum, library, and visitor center. Efforts to restore the surrounding grasslands to their original state are currently under way.

Aerial view of Fort Union reconstruction, near Buford, North Dakota —Courtesy of Jim Wark, from *Discovering Lewis and Clark from the Air* by Joseph A. Mussulman and Jim Wark; used by permission

Reconstructed fort, today —Courtesy of National Park Service

Plate XVI. *Fort Union and Distribution of Goods to the Assiniboines.* Artist: Stanley.

Manuscript watercolor of Fort Union by John Mix Stanley —Courtesy of Yale University Art Gallery and used by permission

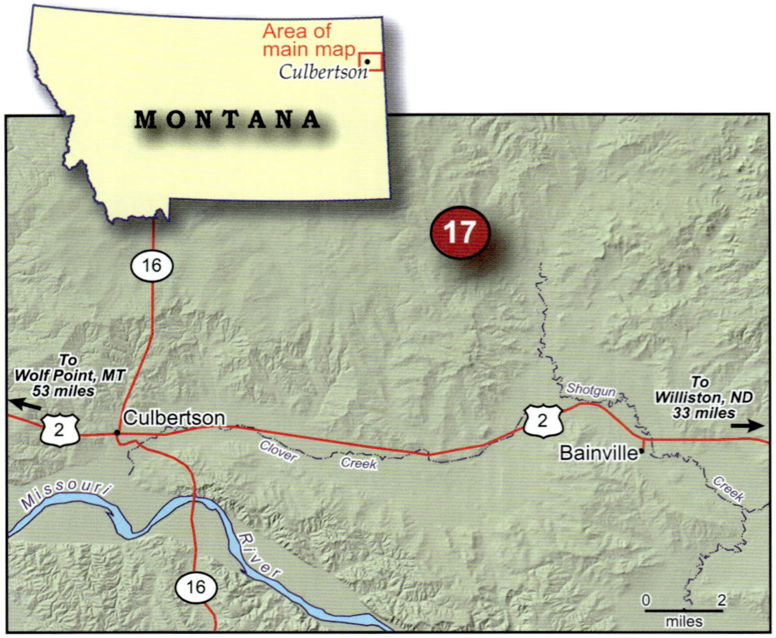

STEVENS AND HIS MEN LEFT FORT UNION on August 9 and proceeded westward along the north side of the Missouri River toward Fort Benton. They were accompanied by Fort Union's post factor Alexander Culbertson and his wife, Natawista. Culbertson was very familiar with the region, especially after he explored and opened a new trail to Fort Benton in November 1851.[49] His route, which followed the Milk River part of the way, avoided the rugged terrain that prevailed along the Missouri River farther south.

About twenty-eight miles west of Fort Union, the survey team met a Blackfeet party under White Man's Horse on August 12. The Blackfeet were camped on the Big Muddy River, having just returned from a horse raid on the Flatheads, who lived west of Rocky Mountains. Stevens briefly described the meeting:

> Before leaving the Big Muddy I had a long conversation, already alluded to, with the White Man's Horse, (see sketch,) the chief of the war party of the Blackfeet. He had frequently visited the Bitter Root valley, and stolen horses from the Flatheads. He observed, "I take the first Flathead horse I come across; it is sure to be a good one." He and one of his men had just returned from the Flathead country, and they gave a very favorable description of the route, assuring me (pointing to my wagons and Pembina carts) that there would be no difficulty in taking them through the mountains.[50]

Stanley's image depicted tribal leaders visiting Stevens's camp. Stevens is standing addressing the group. In the upper left corner of the image, the natives' camp is evident on a small hill. In contrast to his accounts of visits with the Red River Hunters and the Assiniboines before reaching Fort Union, Stevens provided less detailed information about the Blackfeet. It is apparent that Stevens's report was changing in character and detail. When the journey began, he recorded rich and varied descriptions of the Indians he encountered. Near Fort Benton, the governor was becoming impatient and anxious to complete the expedition. Compounding this impatience was his continuing pain, recurring from a rupture he suffered during the Mexican War. As the journey progressed, he walked less and rode horseback or in a wagon more often.

Plate XVII. *Council with White Man's Horse.* Artist: Stanley.

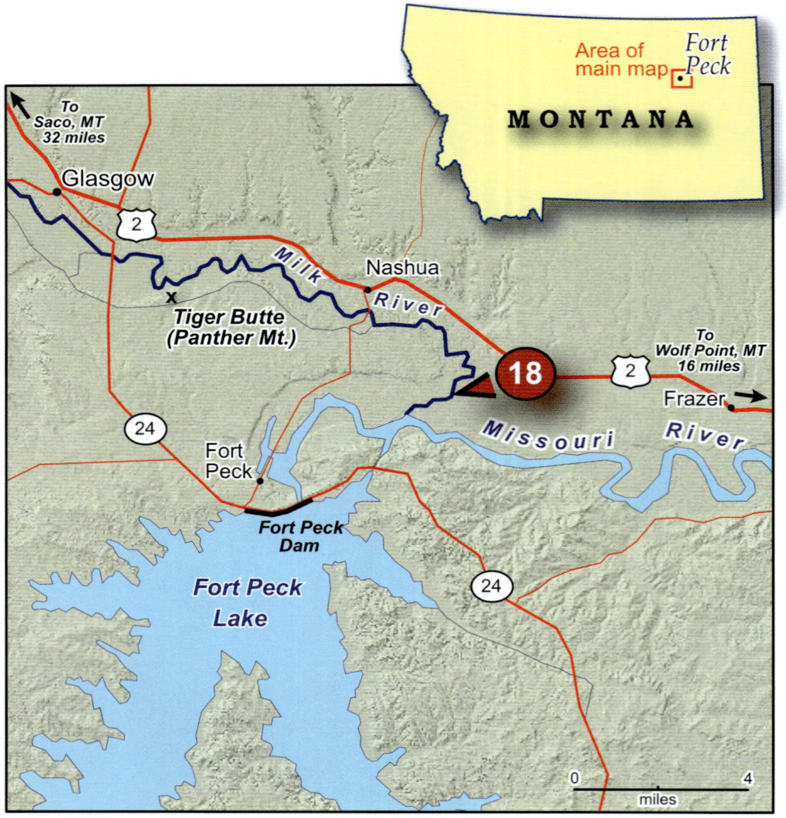

On August 18 the expedition reached the mouth of the Milk River and camped in a cottonwood grove along its banks. It proved to be a very nice campsite, and the party stayed there several days. Part of the time was spent making charcoal, fuel for the blacksmith. Equipment was also repaired. Stanley took advantage of the extra time to draw. The survey team took astronomical observations.

> Passed through to-day villages of prairie dogs. Crossed the Porcupine river about five miles from camp. Encamped on the Milk river, 16 miles being the day's march. Here we determined to remain a day to prepare charcoal for the blacksmith, and to make observations for the geographical position of its mouth, which is considered a very important point in the survey.[51]

The Milk River is a name bestowed by Meriwether Lewis, who thought the water was "the colour of a cup of tea with the admixture of a tablespoonfull of milk."[52] The stream's color results from its flowing through and eroding soft sedimentary rocks. One key source of this sediment is a gorge in Writing-on-Stone Provincial Park in Canada. The Milk's merger with the Missouri River is dramatic as the former's warm, silty water collides with the rapid-moving, cool, and clear water of the latter.[53]

This is one of the few lithographic plates featuring wildlife as the image's core subject. Small groups of antelope (pronghorns) dominate both the foreground and background. The larger animals in the foreground and smaller ones in the background create an illusion of depth.

Aerial view of confluence of Missouri and Milk Rivers, near Fort Peck, Montana —Courtesy of Jim Wark, from *Discovering Lewis and Clark from the Air* by Joseph A. Mussulman and Jim Wark; used by permission

Plate XVIII. *Milk River near Junction of Missouri.* Artist: Stanley.

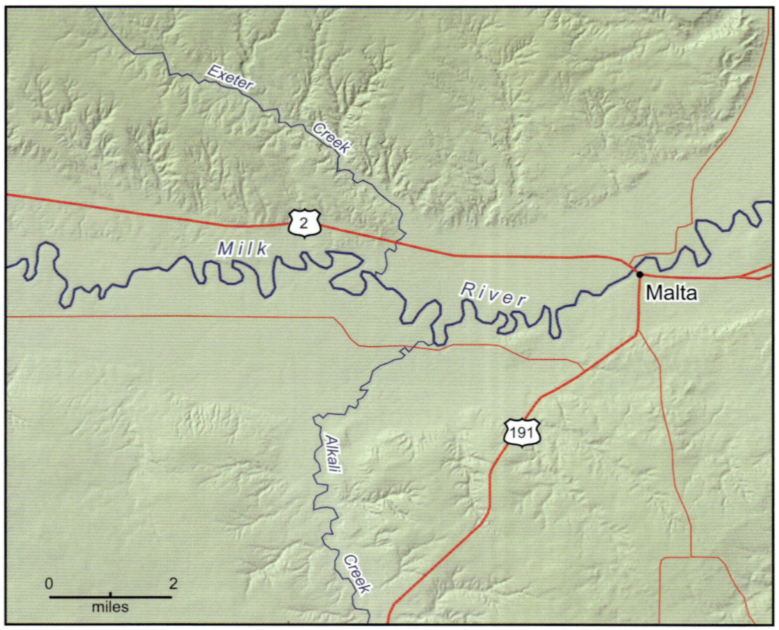

Topographic map of Milk River, illustrating a misfit stream

WHILE CAMPED AT THE MOUTH OF THE MILK RIVER on August 18 and 19, Stanley created an image of the camp. Stevens described the setting:

> Our camp is surrounded by a large grove of cottonwood, (see sketches,) and near it was a delightful spring of water. The valley of Milk river is wide and open, with a very heavy growth of cottonwood as far as the eye can reach.[54]

Some of Stanley's images, such as this one, featured grandiose presentations with trees that had rugged bark and a statuesque form. Considering the fact that the plains were largely devoid of trees, these lofty specimens were very attractive to the artist.

The trees observed here are Great Plains cottonwoods, scientifically known as *Populus deltoides*.[55] This large tree is found from the Great Plains to the eastern border of the Rocky Mountains, extending north into Canada. It is a subspecies of the eastern cottonwood, but unlike its cousin it has hairy buds and lighter yellow twigs. Larger specimens can grow higher than ninety feet, and during the day of steamboat travel along the Missouri, the trees were a significant source of fuel wood.

As Stevens noted, the Milk River occupied a very wide valley. The present-day river is too small to have created this large feature. Geomorphologists identify it as a "misfit stream." As the accompanying topographic map shows, the Milk River has etched its small meandering shape into an incongruently large valley and broad floodplain. This is because the lower valley was originally created, not by the Milk River, but by the erosive action of the Missouri River before continental glaciers forced it southward toward its current location. After the newer channel formed, the original channel was abandoned. After the glaciers retreated, the Milk appropriated a segment of the abandoned channel for its own irregular course.

Plate XIX. *A Cottonwood Grove.* Artist: Stanley.

STEVENS CONTINUED TO MOVE WESTWARD following the meandering course of the Milk River. Not surprisingly, much wildlife was seen on the grassy floodplain and surrounding uplands:

> August 21. This morning was clear, cool, pleasant, and delightful for moving. Engineer parties, both yesterday and to-day, have been actively at work getting in the country bordering the route of the main party. . . . I dispatched a small party across Milk river to Panther Hill (see sketch) to observe the country. Game was very abundant; plenty of buffalo, antelope and beaver. A heavy rain and thunder storm occurred about noon. . . . Wild horses were reported as having been seen to-day by the reconnoitring parties. A fine eagle was shot and brought in to Dr. [George] Suckley, our naturalist.[56]

The Milk River's watershed is still identified as a major wildlife habitat for deer, beaver, mink, and waterfowl such as sandhill cranes and great blue herons. Sadly, many of the most notable inhabitants (the bison, the grizzly bear, and the wolf) have largely disappeared.

The stream's geomorphology is clearly depicted in the foreground of Stanley's image. Notice the gentle bank on one side and the steep banks on the other, created by the stream's flow. Here at the river's bend, the outer shoreline, or cutbank, is being eroded away rapidly, while the inner shore, or slip-off slope, collects sediment deposits.

Milk River near Havre, Montana
—Photograph by Paul McDermott

Plate XX. *Milk River and Panther Mountain.* Artist: Stanley.

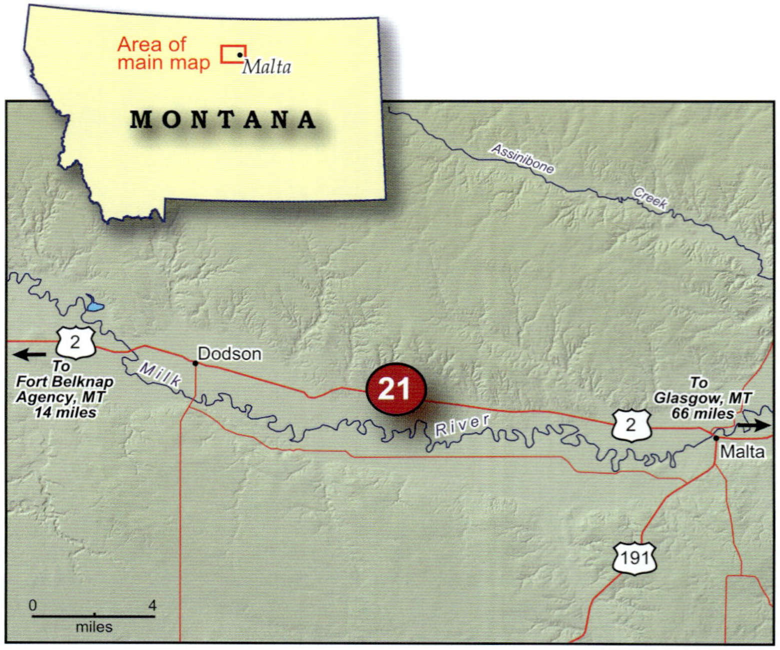

STEVENS FIRST ENCOUNTERED THE GROS VENTRE PEOPLE on the banks of the Milk River on August 25. The camp, situated about seventy miles east of present-day Havre, Montana, comprised 300 lodges, 1,000 or more horses, and about 2,000 people. The survey party held a council and feast for the Gros Ventres in the early afternoon of the next day:

> The necessary preparations were made to prepare the feast, and about 1 o'clock the Indians were seated in squads of twenty or thirty around, to the number of nearly two hundred. Before the feast the Indians seemed to be in high glee, passing the time in singing their songs, accompanying them with rattles made of the hoofs of antelopes strung very fancifully upon a piece of wood about a foot long, with which they marked time.[57]

As part of his negotiations, Stevens presented the tribe with a large gift of trade goods. The combined weight of the gifts was approximately one ton; among the items were blankets, shirts, calico, knives, beads, paint, gunpowder, shot, tobacco, and hard bread. The governor hoped to promote amiable relationships between whites and the natives in anticipation of railroad construction and further settlement.

At this location, the party observed a very bright comet in the northwest sky. Stevens and his men were enthralled by this astronomical treat:

> The comet spoken of a day or two ago appeared to-night, shining most brilliantly in the northwestern portion of the heavens, and the aurora borealis presented a beautiful appearance.[58]

Research reveals that this was Comet Klinkerfus.[59] The accompanying celestial view reflects the sky as it appeared on August 25, replicating what the men most likely saw.[60]

Reconstruction of night sky showing Comet Klinkerfus, August 25, 1853 —Courtesy of Dr. William Davis, National Institute of Standards and Technology

Plate XXI. *Distribution of Goods to the Gros Ventres.* Artist: Stanley.

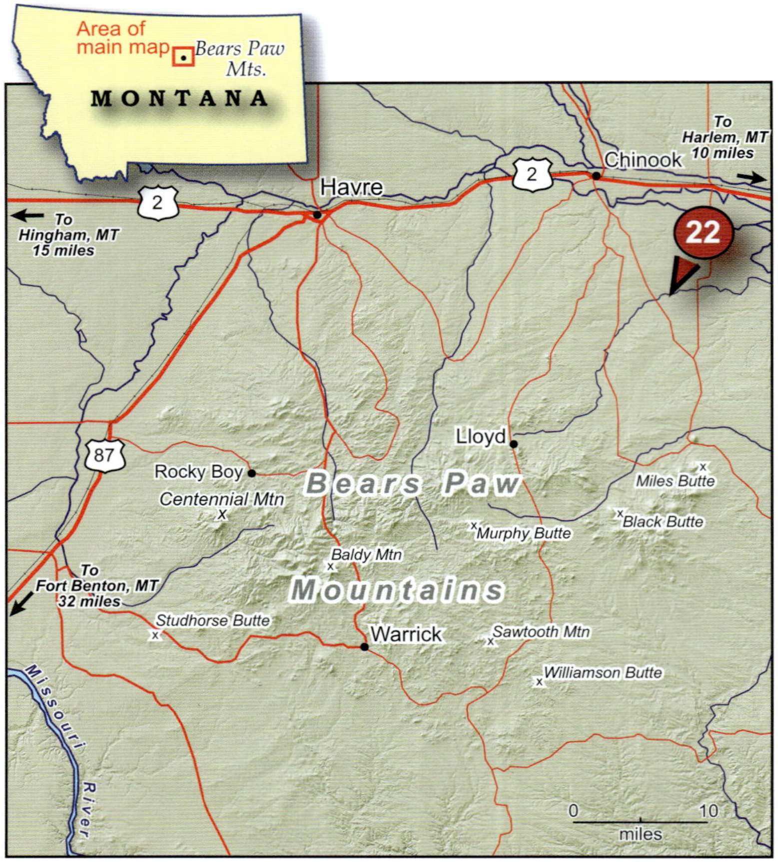

NEAR THE PRESENT-DAY SITE OF **HAVRE, MONTANA,** Stevens left the Milk River behind and turned the expedition southwest toward Fort Benton. On August 30, Stevens recorded their progress:

> Yesterday, we were in sight, of the Bear's Paw, quite a broad and rugged mountain upheaval, stretching from Milk river to the Missouri. (see sketch.) . . . After leaving Milk river we crossed several branches of Box Elder creek, which is a tributary of Milk river that has its source very near the Missouri, and is on our general line to Fort Benton.[61]

In this illustration, Stanley was overlooking the Milk River. In the background is the faint outline of peaks, a segment of the Bears Paw Mountains. It is a peaceful scene. Neither humans nor animals were included to create a sense of scale, so the viewer has only the trees to use for comparison. No single object dominates the foreground except a small hill. Like many other images in the series, the foreground is darker than the background to give the perception of depth.

Good artistic representation typically entails dividing the image area into thirds. Vertical and horizontal lines define the image. Where the lines meet, visual power points are created. A thoughtful artist will use these points to arrange the image's subjects. Here, however, Stanley divided the image into equal halves, resulting in too much sky. Stanley mitigated the problem by filling the bald sky with visually interesting clouds. Overall, this is one of the least compelling plates in the series.

Plate XXII. *Milk River—Bear's Paw Mountain in Distance.* Artist: Stanley.

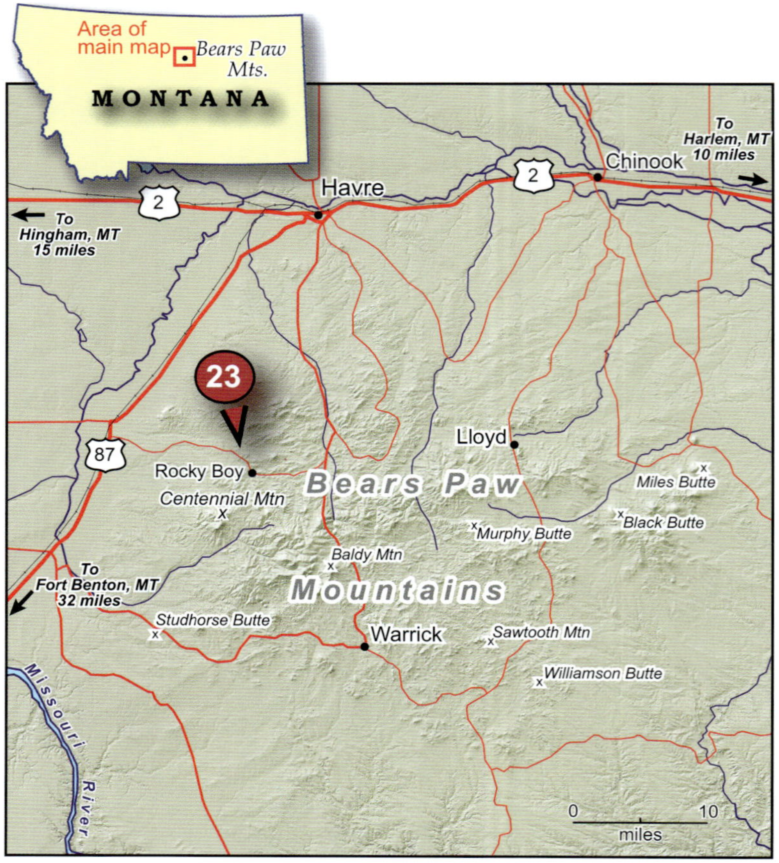

Bears Paw as viewed from the west, south of Havre, Montana
—Photograph by Paul McDermott

As they neared the Bears Paw Mountains, Stevens sent three men to investigate the mountain complex to the south. Stanley made sketches of the mountains while the other two took measurements. As part of their duties, Stevens ordered them to ascend the highest peaks:

> [August 30.] I sent off Lieutenant Grover, Mr. [F. W.] Lander and Mr. Stanley to make an examination of the Bear's Paw, as far as it could be done by ascending one of its highest peaks, and also to sketch in the whole surrounding country, especially the portion between the Bear's Paw and the Rocky mountains.[62]

Most likely they climbed Centennial (6,074 feet) or Baldy (6,916 feet) Mountain. The accompanying map shows an overview of the range. As Stevens viewed the landform from afar, he noted both its practical and aesthetic features:

> From the best information which I can gain of these mountains, they are intercepted with valleys which give easy passes for pack animals, and probably no difficulty would be found getting through with wagons. . . . The view this afternoon was delightful. Bear's Paw itself presents a rugged, grotesque appearance, and it requires no great stretch of the imagination to see in it the paw of a grizzly bear, ready to spring upon the plain. (see sketch.)[63]

Stanley's view shows the small group approaching the Bears Paw. Each peak has a rough conical shape with its top beveled off. On the image's right side, a volcanic dike protrudes up toward the sky. Although artists usually avoid center placement of a focal point, Stanley has included a small scrub directly in the lower center, probably to provide a sense of scale and depth.

When the expedition reached this part of Montana, they started encountering igneous rock formations, of which the Bears Paw was the first.[64] From this point on, they encountered other similar formations, especially between the Sun River and the Dearborn River. Igneous laccoliths and dikes dot the landscape throughout this region. One especially attractive example is Birdtail Rock, or Birdtail Butte (see photo on p. 88 and drawing on p. 172).

Plate XXIII. *Bear's Paw.* Artist: Stanley.

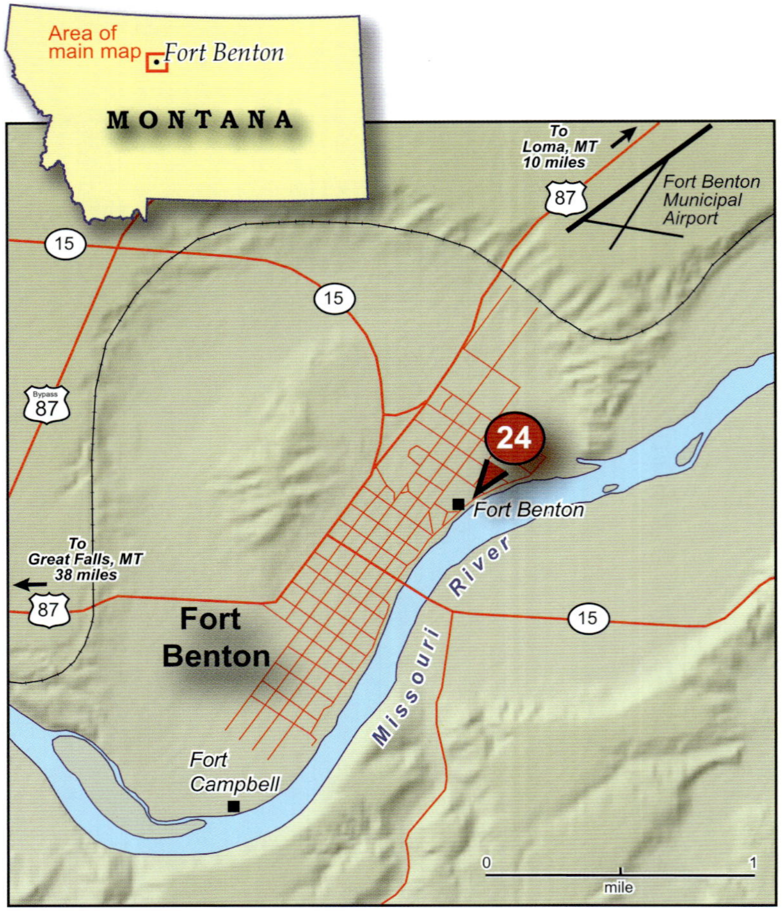

STEVENS AND HIS PARTY REACHED FORT BENTON, situated on the northwest bank of the Missouri River, on September 6. The men had traveled for ninety days and covered 1,092 miles.[65] From the post, Stevens sent reconnaissance parties into the surrounding region, especially toward the Rocky Mountains to the west. The main party rested here for ten days. Stevens briefly described the fort in his report:

> Fort Benton is smaller than Fort Union. Its front is made of wood and the other sides of adobe, or unburned brick. It usually contains about a dozen men, and the families of several of them. The Blackfoot Indians are the principal traders here.[66]

During his stay at Fort Benton, Stevens held council with the Blackfeet. He observed the importance they gave to dress:

> The chiefs and warriors were all richly caparisoned. Their dresses of softly prepared skins of deer, elk, or antelope were elegantly ornamented with bead-work. These are made by their women, and some must have occupied many months in making. The other articles of their costume were leggings made of buffalo skins, and moccasins, also embroidered, and a breech-cloth of blue cloth. Their arms were the Northwest guns, and bows and arrows. On all solemn occasions, when I met the Indians on my route, they were arrayed with the utmost care. My duties in the field did not allow the same attention on my part, and the Indians sometimes complained

Fort Benton as it appeared in 1860 —Courtesy of Fort Benton Historical Association and used by permission

Reconstructed interior of Fort Benton —Photograph by Paul McDermott

of this, saying, "We dress up to receive you, and why do you not wear the dress of a chief?"[67]

Today, Fort Benton is a quaint town of 1,500 people. It marks the head of navigation on the Missouri River—one local history refers to Fort Benton as the "World's Innermost Port."[68] It began as a trading center with two forts. Originally known as Fort Clay, the settlement was renamed Fort Benton in honor of prominent Missouri senator Thomas Hart Benton, who played an important role in promoting western exploration.

In Stanley's image, we see the fort complex as it appeared in 1853. The scene has an unusually tranquil appearance. Horses graze quietly along the riverbank under the watchful eye of the natives. The main adobe complex dominates the image; beside the fort are a few very prominent tepees, whose size appears to be radically out of proportion to the fort. In the background and farther upstream (south-southwest) is the outline of a smaller fort.

The scene's quiet feeling may indicate that Stanley used a daguerreotype in preparing the sketch. Stevens acknowledged that Stanley took photographs at the fort:

> Mr. Stanley commenced taking daguerreotypes of the Indians with his apparatus. They were delighted and astonished to see their likenesses produced by the direct action of the sun. They worship the sun, and they considered that Mr. Stanley was inspired by their [own] divinity, and he thus became in their eyes a great medicine man.[69]

Notice the small bateaux tied up on the riverbanks adjacent to the fort. Significant transportation changes were coming. In 1860, when the first steamboats began arriving, the community expanded rapidly. Later, Fort Benton became the terminus of Lt. John Mullan's military wagon road, constructed between 1859 and 1862. The road project, which originated at Fort Walla Walla, about 624 miles to the west, was completed to Fort Benton in 1860, with further improvements taking place in 1861 and 1862.

Plate XXIV. *Fort Benton.* Artist: Stanley.

Manuscript watercolor of Fort Benton by John Mix Stanley —Courtesy of Yale University Art Gallery and used by permission

WHILE THE SURVEY PARTY WAS ENCAMPED at Fort Benton during the first half of September, Stevens sent out a number of small exploratory parties. His objective was to obtain geographical information for a broad region along the proposed railroad route and to determine the most practical passes through the Rocky Mountains. One party, led by Lt. Cuvier Grover, was directed to the Bitterroot Valley to rendezvous with Lt. Rufus Saxton's supply column. Saxton had accompanied Capt. George B. McClellan to Puget Sound and was now bringing supplies east to Stevens's party.

At the same time, Stevens himself led a small party north to Cypress Mountain, Canada. Several days out of Fort Benton, he received a dispatch from Grover reporting that he had met with Saxton in the Bitterroots. Information in the dispatch made Stevens change his plans:

> Lieutenant Grover intimated in his brief letter that Lieutenant Saxton reported the route could not be traversed by wagons. This changed the aspect of affairs, and I determined to send Mr. Stanley to the Piegan camp with the interpreter Hammell, and to return immediately with Mr. Culbertson to Fort Benton. I determined also to defer the examination of the Marias Pass to another season.[70]

Stanley's depiction of the encounter between Stevens and Grover's courier focused on the men rather than the landscape. Although none of the men are identified, the figures in the image center may be Little Dog, a Blackfeet chief accompanying the party; Stevens; and the courier, Baptiste Champagne. Other members of the party included "Mr. Culbertson, special agent, Mr. Stanley, artist, August Hammell, interpreter, and three voyaguers."[71]

One interesting aspect of the image is that the horses have facial expressions that are almost human. While Stanley is noted for this proclivity, it is not certain why. Was he trying to be humorous, was he unable to render horses well, or was this an artist signature?

Plate XXV. *Lieut. Grover's Despatch—Return of Governor Stevens to Fort Benton.* Artist: Stanley.

THE MARIAS IS KNOWN AS THE "RIVER OF DECISION." When the Lewis and Clark expedition visited the junction of the Marias and the Missouri, they were faced with an important decision. Which of the two rivers was the real course of the Missouri? They chose the southern fork, which proved to be the correct choice.

Stevens's party reached the Marias on September 1. It had taken nearly three months to arrive at this point. Stevens described the area:

> The descent to this river on the trail is somewhat steep, the prairie plateau being some two hundred and odd feet above the river bottom. The river itself here presents a beautiful view. It is a clear, limpid stream, flowing over a pebbly and sandy bed, the bottoms lined with cottonwood of heavy growth, with thickets of the service and other berries.[72]

Members of the expedition crossed various portions of the Marias during the first few weeks of September 1853. Engineer A. W. Tinkham crossed the river about seventy miles from its mouth on September 8. Stanley and Stevens crossed it on their journey north toward Canada on September 10. The portion of the Marias illustrated in the report was probably at Stevens's crossing, about thirty miles upstream from its juncture with the Missouri:

> The stream at our crossing was about fifty yards wide, one foot deep, and of somewhat rapid current, and the river valley was about a mile wide. There was plenty of cottonwood.[73]

Confluence of Marias and Missouri Rivers —Photograph by Paul McDermott

Plate XXVI. *Marias River.* Artist: Stanley.

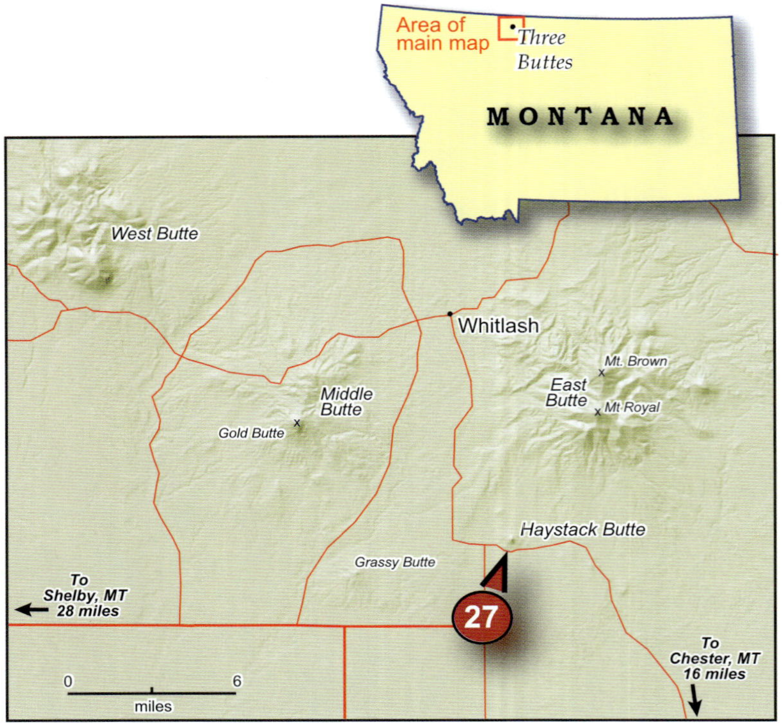

ABOUT EIGHTY MILES WEST-NORTHWEST of Fort Benton is Three Buttes, today known as the Sweetgrass Hills, an important Native American hunting and religious site. The Blackfeet referred to the region as *katoysix*—"place of sweet pine."[74] Three igneous intrusions, called West, Gold, and East Buttes, define the location. The Blackfeet and other Plains tribes identified this as one of their sacred places, where they sought visions and performed various ceremonies. Stevens remarked on the tribes' use of the site:

> Assiniboines, Crows, and Blackfeet, have marked their summits with monumental heaps of stone, and retained their lodges, in which war parties have awaited the favorable moment to pounce down upon the unguarded wanderer in the plains below.[75]

The place was also significant to the natives because the region's good grazing grasses attracted buffalo herds. It remained a prime hunting ground until buffalo numbers declined. Today the region reveals economic development vastly different from a hunting and gathering society. It is a place filled with oil wells, ranches, and wheat farms.

Stanley's dramatic image shows a Blackfeet buffalo hunt. The Blackfeet were a formidable people living east of the Rocky Mountains, around the Upper Missouri River. In the mid-nineteenth century, they were subdivided into three groups—Siksika, Kainah, and Piegan.[76] The Blackfeet were excellent horsemen, as shown in Stanley's action-filled illustration. The sketch, which was probably made on September 17 or 18, features Blackfeet hunters on horseback attacking a panicked herd. Behind them, Haystack Butte, an outlier, protrudes in front of the broad expanse of the higher East Butte. Interestingly, in comparing the modern scene with Stanley's image, one notices that the background has apparently been altered artistically, and the scene appears somewhat impressionistic.

Haystack Butte at Three Buttes
—Photograph by Paul McDermott

Plate XXVII. *Blackfeet Indians—Three Buttes.* Artist: Stanley

THE MAIN EXPEDITION DEPARTED FROM FORT BENTON on September 16. Stevens made his departure six days later. His comments upon leaving the fort bordered on sentimental:

> As we left the fort, soon after sunrise, we felt that our short sojourn there had only let us into the experiences and vicissitudes of life in a remote mountain post, surrounded by numerous Indians, and accessible to information from home but once or twice a year, but we felt also that we had made warm and fast friends of all the inhabitants of that region—voyageurs, Indians, and gentlemen of the Fur Company. And I ascribed it mainly to one reason, that they felt the beneficent results would flow from the expedition, and they saw that every man was thoroughly in earnest in furthering it.
>
> In order to make a long march this day, the evening before I despatched my train to a point up the Teton, some twelve miles from Fort Benton; and there Mr. [Isaac F.] Osgood and Mr. Stanley, who had remained behind with me at Fort Benton, and myself breakfasted with the rest of our party. The accompanying view gives some idea of the valley of the Teton.[77]

Stanley's view of the Teton Valley depicts a relatively wide valley floor filled with clustered trees and scrubs. Smaller streams have eroded steep courses into the valley's angular slopes.

The Teton River flows 160 miles eastward from the Rocky Mountains to join the Marias River a few miles upstream from where the Marias joins the Missouri, northeast of Fort Benton.

Teton River Valley near Fort Benton —Photograph by Paul McDermott

Plate XXVIII. *Teton Valley*. Artist: Stanley.

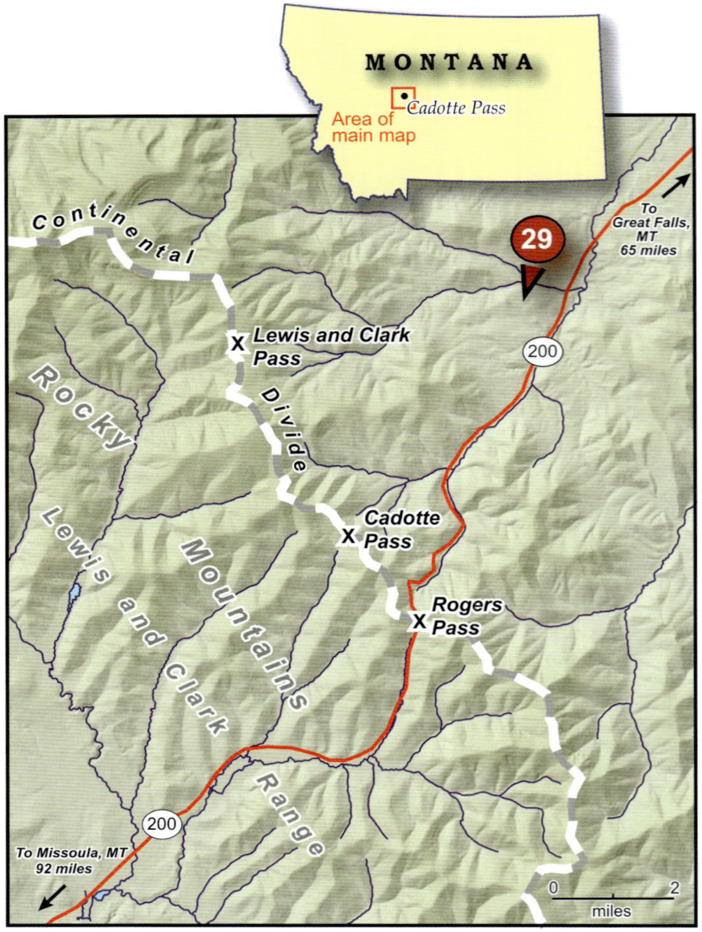

Leaving the Teton River Valley, the main survey party traveled southwest toward the Rocky Mountains. The members crossed the Sun River, passing by Crown and Square Buttes, and then traveled through a region of unusual igneous formations protruding above the plains, which Stanley depicted in this illustration. One of the most prominent of these is Birdtail Butte, which is just southeast of Montana state route 200 near the boundary of Cascade and Lewis and Clark Counties. This butte was a featured landmark in the maps and sketches documenting the construction of John Mullan's military wagon road (see drawing on p. 172). Stevens passed it on September 23, 1853.

At this juncture, the party followed an easy egress provided by the Dearborn River, which flowed northeastward from the mountains. Upon reaching the main mountains, Stevens had the choice of three passes by which to cross the Continental Divide. These passes, which provide access from the Dearborn to the Big Blackfoot River systems, are known today as Lewis and Clark, Cadotte, and Rogers Passes. He chose Cadotte Pass, which was also a popular route for Native Americans going to buffalo hunting grounds on the high plains.

When they crossed the pass on September 24, the weather was rainy, and the men were cold:

> As we ascended the divide, a severe pelting hail and rain storm, accompanied with high wind, thunder, and lighting, suddenly came upon us, and did not abate until we had reached the summit. The wind blew very violently, and the mist resulting from the storm prevented our getting a very clear view of the country before us. It was with great gratification that we now left the plains of the Missouri to enter upon the country watered by the Columbia.[78]

Birdtail Butte, an igneous intrusion near the Dearborn River
—Photograph by Paul McDermott

Plate XXIX. *Approach to Cadotte's Pass.* Artist: Stanley.

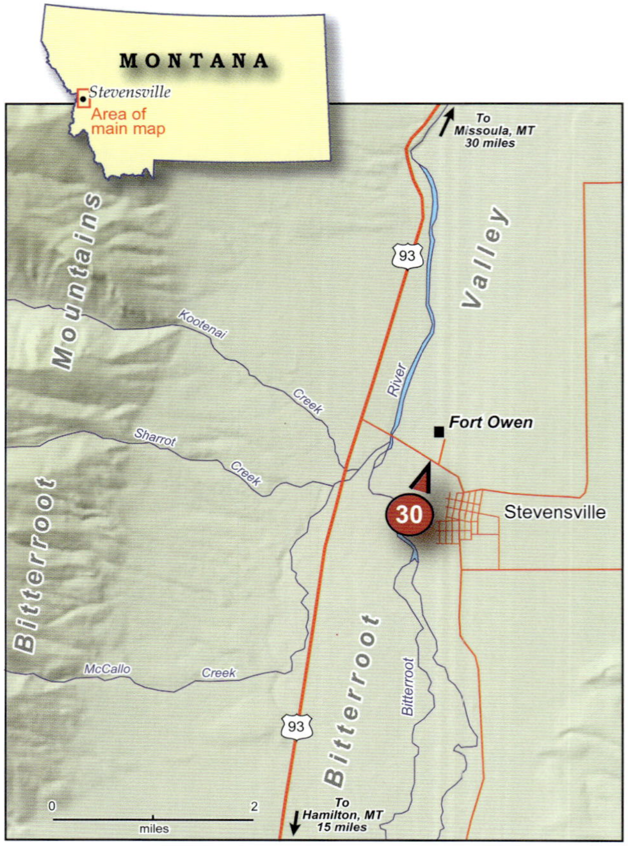

Plan of Fort Owen —Drawing by Philip Mobley

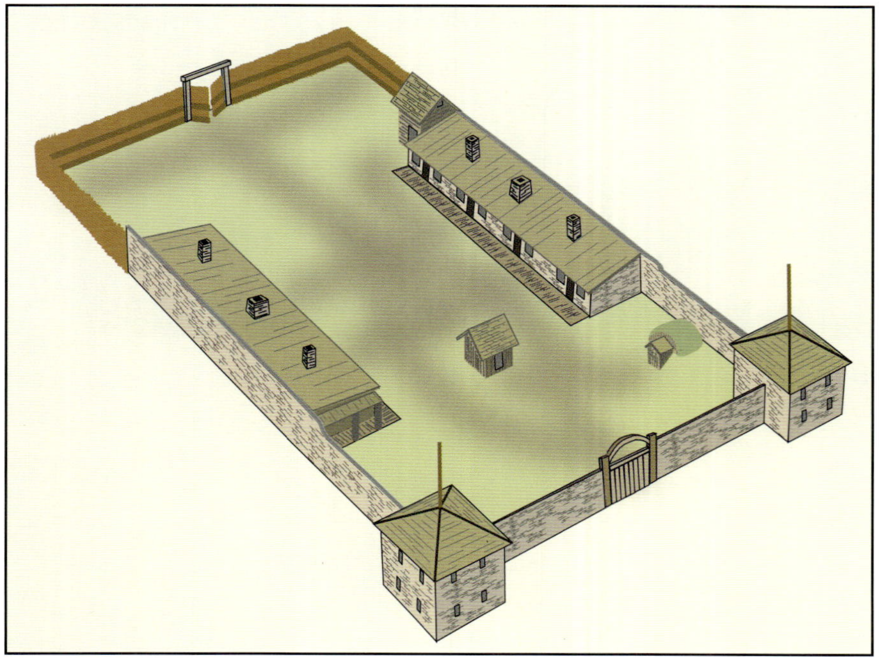

After crossing the Continental Divide through Cadotte Pass, Stevens traveled down the Big Blackfoot and Clark Fork Rivers to the Hell Gate Ronde. Then he headed south along the St. Mary's River (today's Bitterroot) to Fort Owen. He remained at the fort from September 29 to October 3, where he made preparations for the next portion of his journey and met with various Flathead chiefs.

Stevens also decided to leave a small party at Fort Owen to make observations in the valley through the winter. Stevens's son and biographer, Hazard Stevens, explained his father's purpose:

> The governor decided to establish a post in the Bitter Root valley for the winter, under the charge of Lieutenant Mullan, in order to determine the winter regimen of the mountains, the depth and duration of snow, the climate, etc. Thirteen men were left with Lieutenant Mullan, and a large band of animals and ample supplies, and he was instructed to make careful meteorological observations during the winter, to continue the exploration of the mountain section, extending it to Fort Hall on the south, and as far as the Flathead Lake or Clark's Fork on the north, and to keep a watchful and protective eye over the Flathead Indians.[79]

Stanley's image depicted a post complex containing seventeen buildings spread out over a fairly large area. Many of these were outbuildings used as dwellings by employees and some natives. The main fort is pictured, but with very little detail and relegated to the background. In this view, which looks toward the north, the Bitterroot Mountains are on the left and portions of the Sapphire Range are on the right.

Today Fort Owen is a state park, about twenty-five miles south of Missoula, Montana. In 1841 Father Pierre DeSmet established a Catholic mission here. Ten years later, the mission was abandoned and the site was sold to John Owen, who operated it as a trading post.

Plate XXX. *Fort Owen—Flathead Village.* Artist: Stanley.

After leaving Fort Owen, Stevens camped several days in the Hell Gate Ronde. (*Rond* is a French term for a broad, level valley.) The weather had become cooler. Although he was aware that winter was approaching, Stevens was still concerned about promoting good relations with Native Americans in the region. During his short stay at Fort Owen and Hell Gate, Stevens met with Flathead tribal chiefs several times.

Of special concern was Victor, a major chief of the Flatheads, who used the Hell Gate Ronde for grazing his cattle herd. On October 5, Stevens met with Victor at his campsite, which was three miles east of the junction of the Bitter Root (today's Clark Fork) and St. Mary's (today's Bitterroot) Rivers.[80]

The meeting was cordial and included discussions about tribal life and the region's geography. Victor's encampment was small, as attested by Stanley's image. A trail leads toward several tepees including a military tent, which presumably Stevens used during his visit. Around the camp, cattle are resting or grazing.

Victor's people were distributed in a roughly linear swath in the intermontane regions from Flathead Lake to the Bitterroot Valley. A number of Flatheads also congregated at Fort Owen and most likely at Cantonment Stevens as well. They were a partially agricultural people, but like many natives in the region, in the winter they went to hunt buffalo east of the Rockies. As a representative of the government, Stevens was concerned that the Flatheads' conversion to a farming culture was too slow. He told the chiefs, "The Great Father wishes you to raise more wheat and potatoes."[81]

Hell Gate on the Clark Fork River, viewed from the northwestern suburbs of Missoula—Photograph by Paul McDermott

Plate XXXI. *Victor's Camp—Hell Gate Ronde.* Artist: Stanley.

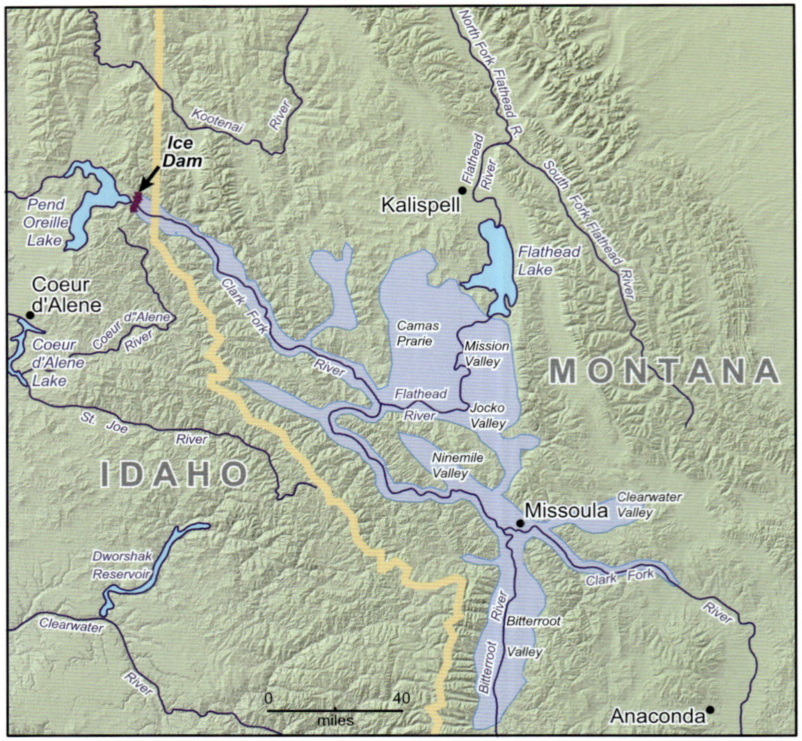

Glacial Lake Missoula

ONE OF ENGINEER A. W. TINKHAM'S ASSIGNMENTS was conducting explorations adjacent to the main route that Stevens followed. Some of his work took place in the intermontane region north of the Clark Fork River and along the western slopes of the Rockies. What is truly amazing is the fact that men like Tinkham frequently explored the wilderness alone. If sick or injured, they were without assistance. They did their work in face of hardships and mortal danger. Yet most survived. The self-reliance, competence, and physical and mental strength of such men were extraordinary.

This illustration, based on a Stanley sketch, shows men relaxing around a campfire. It was October 6, around 5:00 p.m. The men were awaiting their dinner of roasted beef, which was being prepared by Sgt. William Simpson.

Tinkham returned from one of his side trips that evening to rejoin Stevens, and Lieutenant Mullan was paying a visit from Fort Owen. The governor observed the scene:

> Bending over the fire, with huge drops of perspiration rolling from his glowing red face, a picture was presented which Mr. Stanley thought not unworthy a trial of his pencil, while Osgood jokingly told Simpson he was working then for "two dollars a day and roast beef." The meat was cooked in the nicest manner, and at half past five o'clock we sat down to it, having as guests Mr. Tinkham and his party, the returned "lost sheep of the house of Israel," and also Lieutenant Mullan, who arrived in season to join in our meal.[82]

This scene was in the Hell Gate Ronde, very close to where Governor Stevens would hold treaty negotiations with the Flathead in July 1855. In the illustration, on the mountain slopes in the background lay numerous parallel terraces, faintly depicted, documenting the former existence of the huge glacial lake known as Lake Missoula. The ancient lake was the product of meltwater produced from glacial ablation. While the water was trapped in the basin, wave action created small terraces on the hillside. At the same time, sediments formed on the lake bottom; these are known as varves.

Lake varves created by Glacial Lake Missoula
—Photograph by Paul McDermott

Plate XXXII. *Awaiting the Return of Mr. Tinkham.* Artist: Stanley.

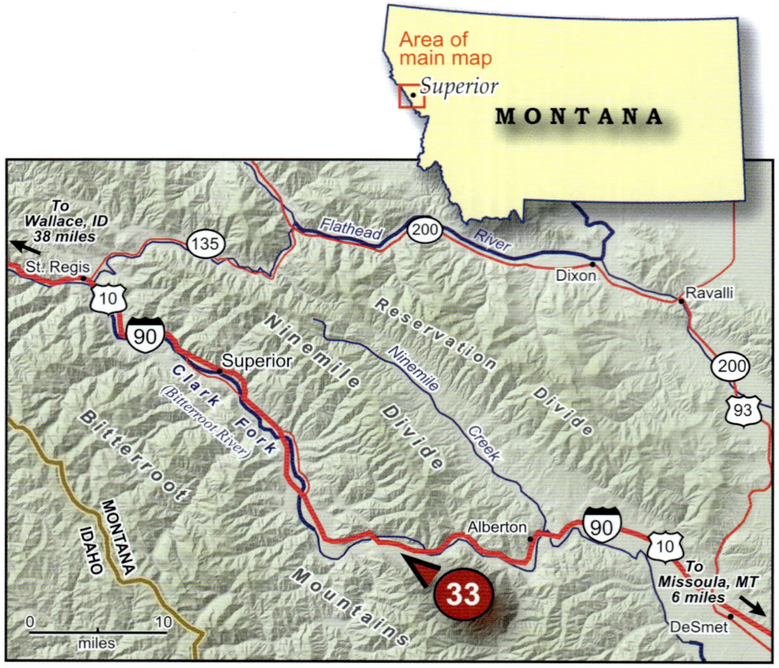

After reaching the Hell Gate Ronde, Stevens ordered Lt. A. J. Donelson to lead his main party north and westward following the Jocko and Flathead Rivers to the Clark Fork. Stevens himself, accompanied by a smaller party, took a more southerly route following the Clark Fork River, known to Stevens as the Bitter Root. On October 8 Stevens encountered a small contingent of Nez Perces moving east along the north bank of the Clark Fork. The survey party was about thirty-nine miles downstream from the Hell Gate Ronde camp, probably meeting the Nez Perces just south of today's Superior, Montana. The governor described the encounter:

> Here we met a band of fifty Nez Perce Indians going to hunt. They have from 250 to 300 horses, most of them splendid animals, in fine condition, and with perfectly sound backs. Women and children helped to compose the band, and babies of fifteen months old, packed in a sitting posture, rode along without fear, grasping the reins with their tiny hands. We met them in the entrance to a narrow place, a mile in length, leading along the water's edge.[83]

The Nez Perces were a large tribe, named by French fur trappers and traders for their pierced noses, though in fact only a few tribal members wore nose ornaments. The Nez Perce territory was distributed over a large region that included parts of what are now northeastern Oregon, southeastern Washington, central Idaho, and far western Montana, including portions of the Bitterroot Mountains.

For food, the Nez Perces gathered assorted plants, including camas roots; fished for salmon and trout; and hunted grazing animals like deer and buffalo. After acquiring horses in the eighteenth century, they became excellent horsemen and buffalo hunters.

The destination of the band Stevens met was the buffalo country on the eastern flanks of the Rockies. The animals gathered there for the grass that was exposed by periodic chinooks, warm, easterly moving winds descending off the mountains.

Plate XXXIII. *Nez Perces.* Artist: Stanley.

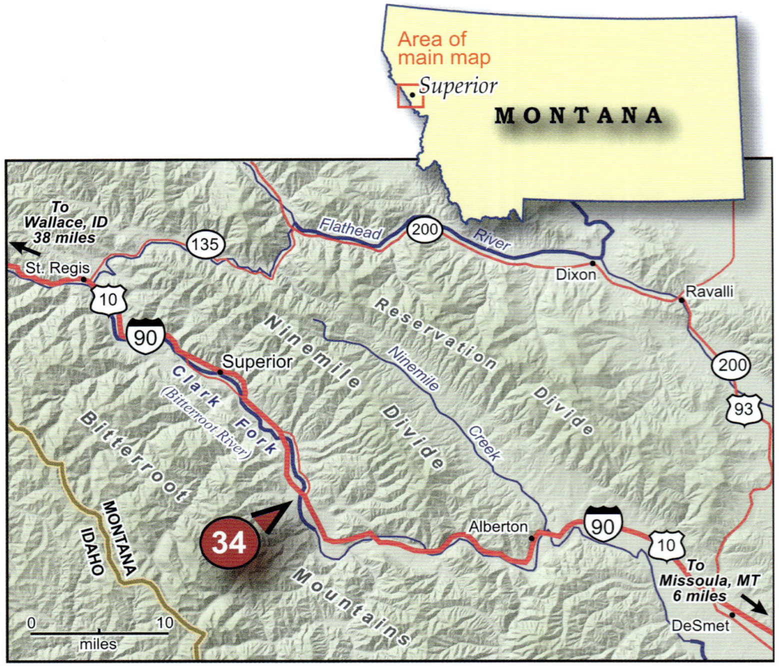

STEVENS'S PARTY FOLLOWED THE BITTER ROOT (today's Clark Fork) River northwestward through a long, linear valley defined by numerous terraces and mountain spurs. Most of this area was once part of ancient Lake Missoula. The group was forced to cross the Bitter Root several times to locate an easily traversed terrace. One of these crossings, which occurred on October 9, was depicted in Stanley's illustration. Stevens recorded the events of this day:

> We started at 8 a.m., and crossed the ford. (see sketch.) The ride of to-day has been rather tedious. We left the valley to get rid of the undergrowth of bushes, and took a trail over the side-hill, which carried us up and down hill successively, and in some instances through woods, occasionally obstructed by fallen timber. At noon we halted at a creek, where we found a single Indian family drying venison.[84]

The river Stevens called the Bitter Root is part of what is now known as the Clark Fork, while today's Bitterroot River, south of Missoula, was then called St. Mary's River. The Clark Fork is about 300 miles long. It begins near Butte, Montana, and flows generally northwestward, twisting and meandering until it reaches Pend Oreille Lake, which feeds the river of the same name. The waters of Pend Oreille River enter into the Columbia River near the Canadian border.

View of the Clark Fork River
—Photograph by Paul McDermott

Plate XXXIV. *Crossing the Bitter Root.* Artist: Stanley.

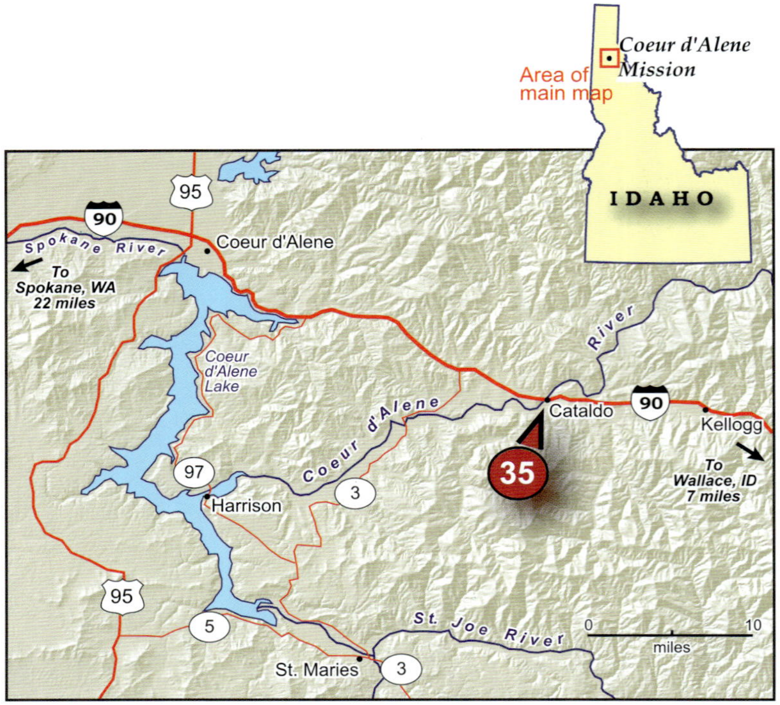

STEVENS'S SURVEY PARTY REACHED the Coeur d'Alene Mission in present Cataldo, Idaho, on October 12. The governor provided a very detailed description of the mission and activities taking place:

> The Mission (see sketch) is beautifully located upon a hill overlooking extensive prairies stretching to the east and west towards the Coeur d'Alene mountains and the Columbia river. About a hundred acres of the Eastern prairie adjoining the Mission are enclosed and under cultivation, furnishing employment to thirty or forty Indians—men, women, and children. . . . The Mission is composed of buildings enclosing a square. Some of them are quite old, but the barn is large and new. The church stands a little distance from the rest, and does much credit to those who erected it. It is constructed upon a plan designed by Père Ravalli, and is of the Roman demi style of architecture. Pulleys and ropes were the only mechanical aids in the construction. . . . The interior is prettily arranged. The altar is supported by two massive timbers of pine which are about four feet in diameter. . . . The priests live in a self-denying manner, and the good effects of their influence over the Indians around them are plainly manifest. There is quite a village of Indians near the Mission. They have some half dozen log-houses, but most live in lodges.[85]

Pencil drawing of Coeur d'Alene Mission by Gustavus Sohon, 1858
—Courtesy of Geography and Map Division, Library of Congress

The next day, he wrote further on mission activities:

> While awaiting the arrival of the train, I was enabled more particularly to observe the manner in which the affairs of the Mission are conducted. Brother Charles has charge of the buildings, and attends to the indoor work, cooks, makes butter and cheese, issues provisions, and pays the Indians for their work, which payment is made in tickets bearing a certain value, "good for so many potatoes, or so much wheat," etc. By this arrangement the Indians are able to procure their subsistence in the summer by hunting and fishing, and have tickets in store for living during the winter. They are well contented, and I was pleased to observe habits of industry growing upon them. In the barn we saw their operations of threshing: four boys rode as many mules abreast in a circle, being followed by two girls with flails, who appeared to be perfectly at home in their business. One half of the barn is reserved for their crops, while the other is arranged for cattle. Their stock at present consists of twenty cows, eight pairs of oxen, and ninety pigs, which are driven to pasture upon the prairie by Indian boys daily.[86]

Exterior view of Coeur d'Alene Mission
—Photograph by Paul McDermott

Stanley's scene is a placid, nearly unpeopled one, looking toward the southeast. In the foreground, an artist, presumably Stanley himself, sits on the stream bank drawing. Stanley often placed himself in his images, perhaps as a way of expressing identity. This graphic signature had the additional advantage of providing a sense of scale.

Another of Stanley's traits is seen on the image's right side. The artist has drawn a small tree standing on the bank. The tree is barely surviving, with only a few branches having leaves. The artist was certainly captivated by oddly shaped vegetation, such as trees with strongly textured bark or plants with dramatic foliage.

Gustavus Sohon also made a rendering of the Coeur d'Alene Mission. His magnificent pencil sketch was done in the fall of 1858, when he visited the site as a member of Col. George Wright's 1858 military expedition.[87] Sohon's was a nearer view, showing the church building and some of the outbuildings in more detail.

Today, the Coeur d'Alene Mission, also known as Sacred Heart Mission or Cataldo Mission, stands quietly adjacent to Interstate 90 as trucks and automobiles go whizzing by. The site is a state park, and the handsome church is recognized as the oldest standing building in Idaho.

Interior view showing mission altar
—Photograph by Paul McDermott

Plate XXXV. *Coeur d'Alene Mission, St. Ignatius River.* Artist: Stanley.

Manuscript watercolor of Coeur d'Alene Mission by John Mix Stanley —Courtesy of Yale University Art Gallery and used by permission

Spokane Falls today —Photograph by Paul McDermott

UPON REACHING THE PLAIN OF THE COEUR D'ALENE RIVER (now called the Spokane River) on his way west to Spokane House, a North West Company trading post established in 1810, Stevens was informed of an interesting double waterfall adjacent to his trail. The cascade, Spokane Falls, stood about nine miles east of the post. He paid a visit to the falls on October 17:

> Leaving camp, Antoine [Stevens's guide], Osgood, Stanley and myself turned from the trail to visit the falls of the Coeur d'Alene river, while Lavatte took the train ahead on the trail to the Spokane House. There are two principal falls (see sketch)—one of 20 feet and the other of from 10 to 12 feet; in the latter, there being a perpendicular fall of seven or eight feet; for a quarter of a mile the descent is rapid, over a rough bed of rocks, and in this distance we estimate a fall of 90 or 100 feet.[88]

Several waterfalls captivated the attention of Stanley and Sohon, including Spokane Falls.[89] This classic image by Stanley shows the falls from a downstream perspective. Standing at its base, he depicted large volumes of water split into two major falls by an island. To add interest, the artist outlined small conifers. These have a delicate appearance and seem struggling to live. Indeed, this is a harsh place. Soil is scarce, and floods frequently threaten their survival.

At these falls, the river actually dropped 150 feet, more than Stevens's estimate. Modern visitors will see that the falls have been tamed, and its bare basaltic terraces are often void of water. The nearby shoreline is now dominated by the city of Spokane, its metropolitan area filling the valley. Instead of thundering waterfalls, we see a dam and adjoining powerhouse. Where once was a barren island is today a serene park housing the remnants of a world's fair held in 1974.

Plate XXXVI. *Falls of the Spokane.* Artist: Stanley.

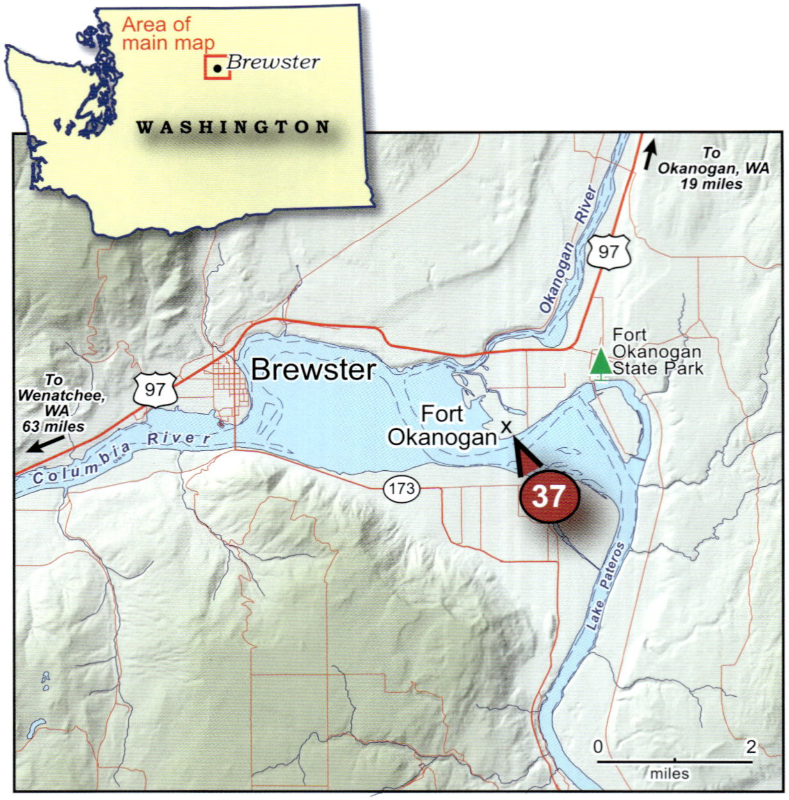

FORT OKINAKANE, TODAY SPELLED OKANOGAN, stood at the junction of the Columbia and Okanogan Rivers. John Jacob Astor's American Fur Company established the fort in 1811. Later it was transferred to the North West Company, then to Hudson's Bay Company. The post's location, on a trail to the Fraser River country in Canada, was commercially strategic.

In this illustration, we see the post as it looked in the fall of 1853. The view was toward the west, with the Cascade Mountains in the background. Even in early October, the range was covered with snow. The main fort stood on a small bluff overlooking the Columbia. Within the complex were four buildings, and another one was situated immediately outside the fort. To the east were four additional log buildings, which were probably used as stables, as suggested by their long, linear shapes. Below the bluff, several large buildings sit beside the river. These appear to be two-story structures, probably used as warehouses.

Although this image is attributed to Stanley, no evidence exists that he personally visited the site in 1853, nor on his earlier tour of the Columbia Plateau in 1847. Capt. George B. McClellan's party had explored this area a few weeks before Stevens arrived, so it is likely that one of McClellan's men supplied the sketch from which Stanley prepared the watercolor and lithograph.

Today, the original sites of several trading posts along the Columbia River, including Fort Okinakane, are underwater. The construction of a dam downstream flooded the site to create a reservoir. However, exhibits at the Fort Okanogan State Park interpretive center explain how the post looked when it was in operation.

Plate XXXVII. *Fort Okinakane.* Artist: Stanley[?]

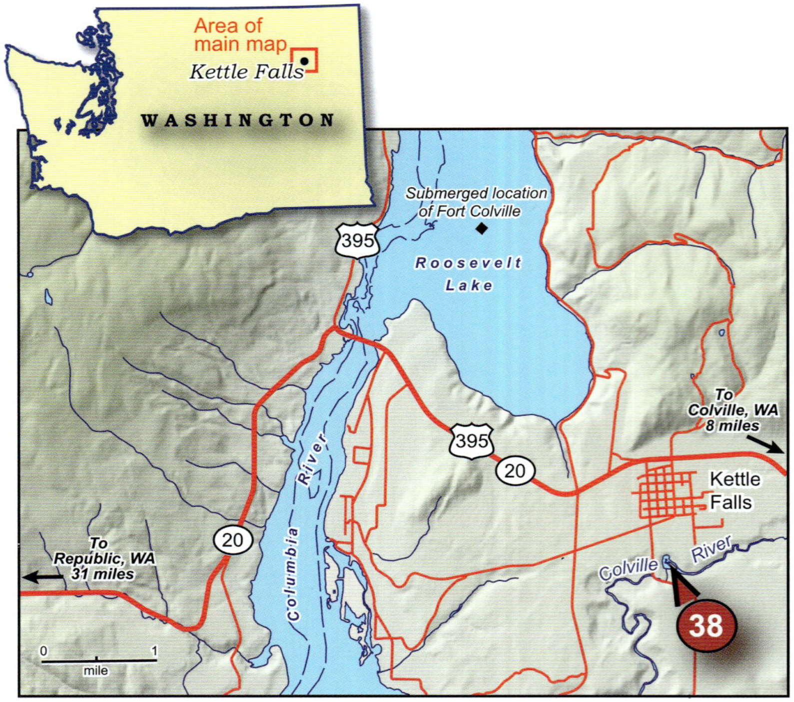

GRISTMILLS WERE EXTREMELY IMPORTANT OPERATIONS in newly settled places. On October 21 Stevens passed by this Hudson's Bay Company mill, which served the region surrounding the trading post of Fort Colville, founded in 1825.[90] The mill was a small operation set along the banks of Mill Creek, a short distance southeast of Kettle Falls. Angus McDonald, the fort's last district manager, ran the mill.

Fort Colville, or *Tshimakain* ("Place of the Springs"), originally dealt in furs, but when the region became part of American territory in 1846, Hudson's Bay Company refocused the fort's business on supplying the needs of settlers, miners, and the military, such as members of the Stevens expedition. The fort also provided a protective presence, particularly for miners who came into the region looking for gold.

It was at Fort Colville that Stevens met up with George B. McClellan's party. McClellan headed the western division of the Pacific Railroad Expedition. His task was to explore the Cascade Mountains and portions of the Columbia River Plateau. The united survey parties stayed several days at the fort, where McDonald showed himself to be a friendly host:

> We then moved off, McDonald presented us with a keg filled with cogniac to cheer the hearts of the members of all parties [Stevens's and McClellan's], and obliged us also to take a supply of port wine. We passed his grist mill on Mill river, the only one in the neighborhood. (see sketch.)[91]

Inside the mill, a set of stones (sometimes called burrs) was used to grind corn or wheat into flour. Examination of Stanley's view reveals that an undershot waterwheel powered the mill.

When Stevens visited Fort Colville in October 1853, Garry, a chief of the Spokane tribe, had recently arrived, bringing wheat from his small farm to be ground into flour. The fact that Garry, whose people lived along the Spokane River plain, came such a long distance indicates that the mill had a large hinterland region.

Plate XXXVIII. *Hudson Bay Mill.* Artist: Stanley.

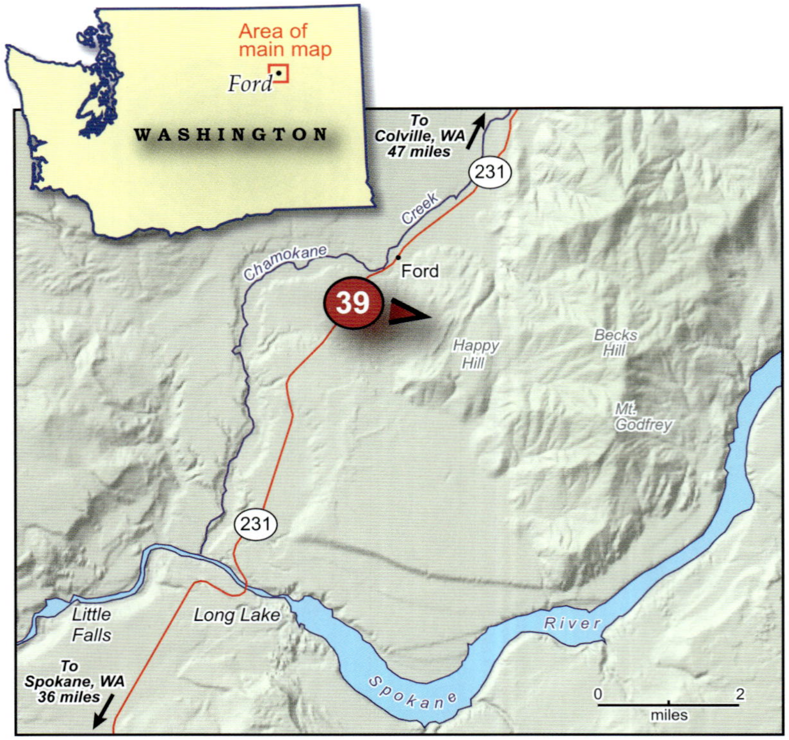

LEAVING FORT COLVILLE, Stevens took his expedition south to Chemakane Mission, located near present-day Ford, Washington. On October 24, he reached the site of this former mission, which had been built by Presbyterian missionaries to serve Native Americans in the region. In 1847, after the Whitman Massacre at Walla Walla, the threat of further hostilities forced the missionaries to abandon the settlement's religious function, though they continued to farm for a few more years. At the time of Stevens's visit, the local natives were still using the land and buildings. Stevens noted the mission's history in his report:

> [A]t noon we reached the old Chemakane Mission, (see sketch,) so called from a spring of that name nearby. The Mission was occupied by Messrs. [Elkanah] Walker and [Cushing] Eel; but, in 1849, in consequence of the Cayuse difficulties, it was abandoned. These gentlemen labored ardently for the good of the Indians. Walker was a good farmer and taught them agriculture, and by them his name is now mentioned with great respect. The house occupied by Walker is still standing, but Eel's has been burned down. The site of the mission is five miles from the Spokane river, in an extensive open valley, well water and very rich.[92]

Stanley's image shows evidence of substantial native activity at the old mission, which appears to have had at least eight buildings. No signs of deterioration are apparent. In the foreground are three native huts that resemble small tents, with a framework of small poles covered by animal skins. A platform that appears to be used for food storage dominates the image's left side.

No visible evidence of the mission remains today. Instead, a bucolic scene traversed by a modern two-lane highway presents a pleasant view for the traveler.

Plate XXXIX. *Chemakane Mission.* Artist: Stanley.

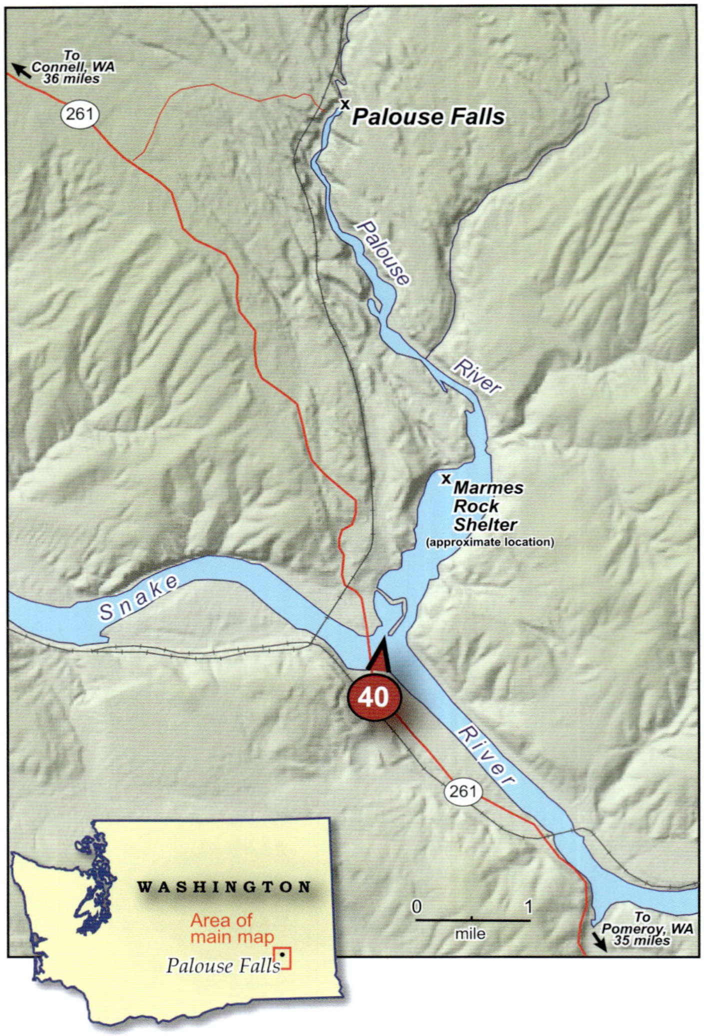

STEVENS'S PARTY REACHED THE JUNCTION of the Palouse and Snake Rivers on November 1. Stevens reported an encounter with the local natives:

> On the Peluse we fell in with two lodges of Indians, and saw a large number of horses belonging to another camp just arrived from up the river. . . . The valley widens considerably for about half a mile from the mouth of the Peluse. The home of the Peluse Indians is near this junction, where they devote much of their time to salmon fishing. The salmon ascend to the falls; but these Indians have a legend which tells of the wickedness of the Indians higher up in the country, and how the Great Spirit, in his displeasure, placed the falls as a barrier to the further ascent of the salmon.[93]

As depicted by Stanley, this was a very appealing place. The view faces north, showing the entrance of the Palouse River into the Snake. Before the last continental glaciations, proto–Palouse Falls was located here. Since that time, the falls gradually retreated upstream six miles.

A short distance upstream on the Palouse is a major archeological site known as the Marmes Rock Shelter. It represents 10,000 years of human occupancy as identified by assorted tools and some skeletal remains.[94] One reason for the human presence at this location was the substantial fishing hole found here, on the Snake River adjacent to its junction with the Palouse. Large sturgeon and other fish became caught in the depression on the river floor, where they could easily be harvested.

Mouth of the Palouse River
—Photograph by Paul McDermott

Plate XL. *Mouth of Peluse River.* Artist: Stanley.

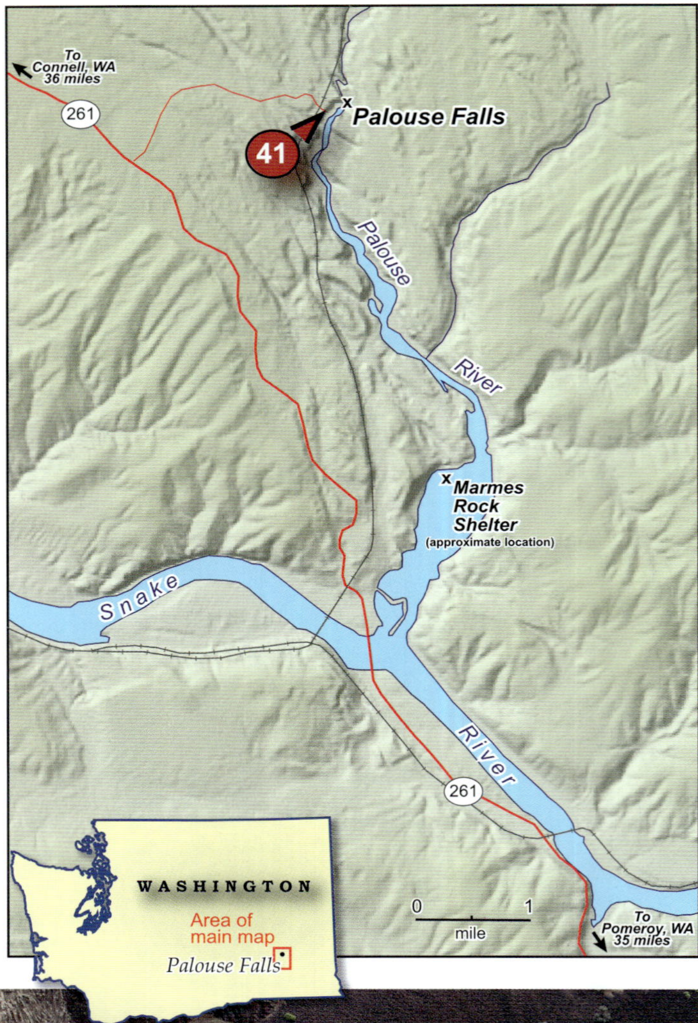

THE STEVENS PARTY DID NOT STOP AT PALOUSE FALLS in 1853. Instead, the governor used Stanley's commentary from the artist's 1847 visit:

> The fall of water, which is about 30 feet wide, cannot be seen from any distant point; for flowing through a fissure in the basaltic rock, portions of which tower above in jagged pinnacles, it suddenly descends some 125 feet into a narrow basin, and thence flows rapidly away through a deep cañon from a point from which the annexed view was sketched.[95]

Stanley's image was made from a vantage point deep within the canyon. It features two figures, apparently fishing on the river's shore. The people and animals in the scene demonstrate the height of the falls. Another captivating component of the landscape here was the spirelike landform seen just left of the falls as the water plunges into the canyon below. Sohon created his own artistic rendition of this natural wonder a few years later.[96]

Stanley's calculation of the falls' plunge at 125 feet was in fact overly conservative. The drop is 198 feet—the equivalent of a twenty-story building. The view is spectacular. From the falls, the Palouse River, which flows over thick basaltic lava flows, zigs and zags for another 6.5 miles until it reaches the Snake River. This waterfall originally formed at the Palouse's intersection with the Snake, but over thousands of years it gradually retreated to its current location.

Palouse Falls today —Photograph by Paul McDermott

Plate XLI. *Peluse Falls.* Artist: Stanley.

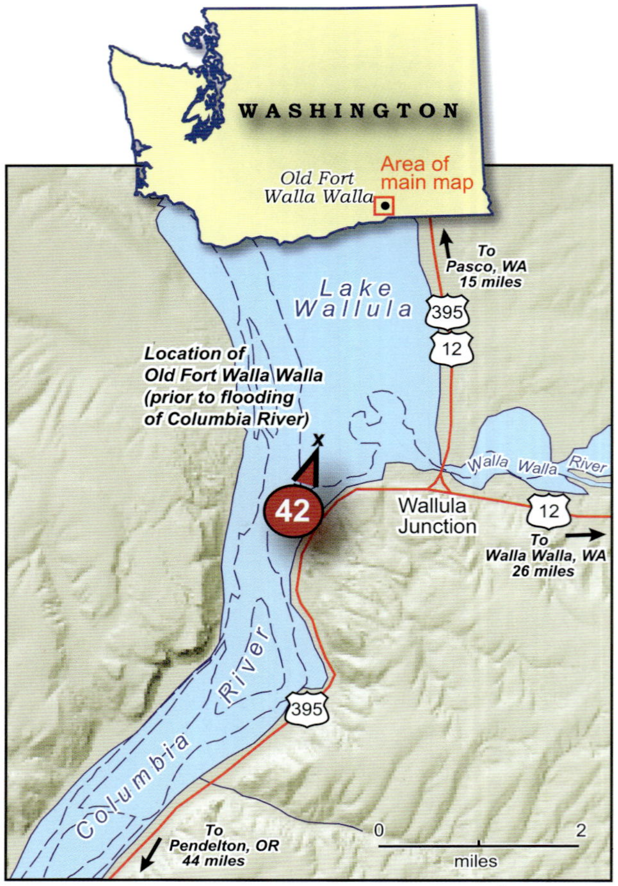

Drawing of Wallula Gap by Gustavus Sohon in 1861
—Courtesy of National Archives and Records Administration

THROUGHOUT THE AUTUMN OF 1853, Stevens was suffering severely from a wound he incurred in the Mexican War. Consequently, he had a difficult time riding a horse. He commented on his condition in his report:

> Although in great suffering, I determined to move with Garry [leader of the Spokane tribe] from Snake river to Fort Walla Walla to-day, leaving Mr. Stanley to come on with my party and train in two days.[97]

The Stevens party reached Fort Walla Walla on November 2, after traveling overland from the Spokane River. From there Stevens explored the Walla Walla Valley, returning to the fort on November 8.

The first Fort Walla Walla, later called Old Fort Walla Walla, was erected as a North West Company fur-trading post at the junction of the Columbia and Walla Walla Rivers, one of the great river bends on the Columbia.[98] Today, the site lies beneath the Columbia's waters in a reservoir behind McNary Dam. Visitors can still enjoy some of the vista seen by early explorers and travelers along the river. When the trading post was abandoned, another Fort Walla Walla, this one an American military post, was built near the present-day city of Walla Walla.

Stanley's view of the post looks south toward the lava-flow bluffs lining the Columbia at the bend. The spot is known as the Wallula Gap. In the foreground, natives are seen conversing in small groups. Several carry firearms—apparently muskets or rifles. In the background, the fort's strong outline projects above the river's bank. The post appears to comprise a few small, indistinguishable buildings and a couple of blockhouses, fortified by a palisade. A larger wooden building marks one entrance through the palisade. A flagpole stands isolated inside the fort.

Gustavus Sohon drew a slightly different sketch at this location in 1861, while participating in the construction of the Mullan Military Road.

Plate XLII. *Old Fort Walla Walla.* Artist: Stanley.

WHILE McCLELLAN AND DONELSON PROCEEDED with their parties overland to The Dalles, Vancouver, and Olympia, where they would reconvene in late November, Stevens took a small party by canoe down the Columbia, between Old Fort Walla Walla and The Dalles:

> I started down the river in a canoe, and reached the Dalles on the 12th. (See sketch.) We took with us two days' provisions, and were four days in reaching the Dalles, having been detained nearly two days in camp by a high wind which blew up the river; but we eked out our scanty stores by the salmon generously furnished us by the Indian bands near us. At the principal rapids I got out and observed the movements of the canoe through them, and, from the best examination which I was able to make, I became at once convinced that the river was probably navigable for steamers; at all events, worthy of being experimentally tested.[99]

Stanley's scene shows the Columbia River in a tranquil state. But a short distance upstream, the river in fact had a turbulent character. For example, at Celilo Falls, twelve miles upstream from The Dalles, the Columbia once plunged violently over flat basaltic rocks. Though a major Indian fishery, Celilo Falls, or Wyam Falls, was a dangerous place, especially for people fishing on the surrounding cliffs and terraces. It was even more treacherous for vessels moving both up- and downstream. Since The Dalles Dam was completed on March 10, 1957, this waterfall no longer exists except for small appendages dotting the river's basin.[100]

Stanley's image captures a small portion of the Columbia River's heritage. Three natives are shown working astride the river. One stands in an elongated boat near the shoreline, using a long pole to maneuver the vessel. Two tents stand nearby, and beside them are drying racks of fish. Faintly in the background, the white outline of Mount Hood emerges through a dramatic cloud formation. The view is generally west-facing through a narrow gap identified as the "Narrows" by Lewis and Clark.

The Columbia is a long river. It begins in the Canadian wilderness and flows through many great bends before reaching the Pacific coast near Astoria, Oregon. From the mountains to the ocean, it travels 1,224 miles and collects water from a basin covering 258,000 square miles.

Plate XLIII. *Dalles.* Artist: Stanley.

Unfinished sketch of Fort Vancouver by Gustavus Sohon, ca. 1854 —Courtesy of Geography and Map Division, Library of Congress

WITH HIS LONG SOJOURN NEARLY COMPLETED, Stevens rested at Fort Vancouver before moving north on foot to Fort Steilacoom on Puget Sound:

> November 15th we went down the river in a canoe, and on the 16th reached Vancouver, (a sketch of which is here given,) where I remained the 17th, 18th, and 19th, as the guest of Colonel Bonneville, and where I also became acquainted with the officers of the Hudson Bay Company.[101]

Fort Vancouver was established as a Hudson's Bay Company trading post in 1825 and remained a viable operation until 1860. Part of its success was its location on the Columbia near the mouth of the Willamette River. The complex was designed to be self-sufficient, developing agriculture, including a substantial orchard north of the fort, and facilities for tanning, shipbuilding, blacksmithing, and other activities. The army post named Columbia Barracks, later changed to Vancouver Barracks, was built adjacent to the trading post in 1849. It functioned as a supply point for military operations around this part of the Columbia River.[102]

Unfortunately, Stevens did not provide any detailed commentary in his narrative about the appearance of the settlement or its activities. However, artist Gustavus Sohon captured considerable detail in this illustration. It is the first Sohon image to be included in the report, and it is one of the few landscape images Sohon drew that showed an important cultural settlement.

In this lithographic view, Sohon viewed Fort Vancouver from the west, looking east up the Columbia River. Energy dominates the scene. People are engaged in various tasks. An empty wagon pulled by two oxen moves toward the foreground. Nearby in the right foreground is another wagon, loaded with four large barrels and pulled by six horses or mules. Three people stand by the roadside. Farther away on the drawing's right side is Fort Vancouver.

Interestingly, the post is not the main feature of the composition. Rather, part of the adjoining settlement, known as "Kanaka," dominates the center and left portion of the image. The village provided housing for workers. Just to the left of the street that runs through the village are several major buildings, the largest of which is the church. On the image's left side, near the horizon, is the army officers' housing, commonly called "Officer's Row."

We have identified two separate images that Sohon created in depicting Fort Vancouver. The most detailed is the manuscript drawing on three-toned paper from which the report's chromolithograph was prepared. The other, from the Library of Congress Geography and Map Division, is an incomplete pencil drawing, also done on toned paper.[103] In the latter composition, Sohon started to draw the settlement complex from a slightly different perspective. From this unfinished sketch, it appears that Sohon changed his mind about how he planned to compose the drawing. The penciled guidelines for the fort provide evidence of his training as a civil engineer. By comparing the two images, we can observe that the artist changed his perspective. The final rendition was drawn from a higher, wider viewpoint.

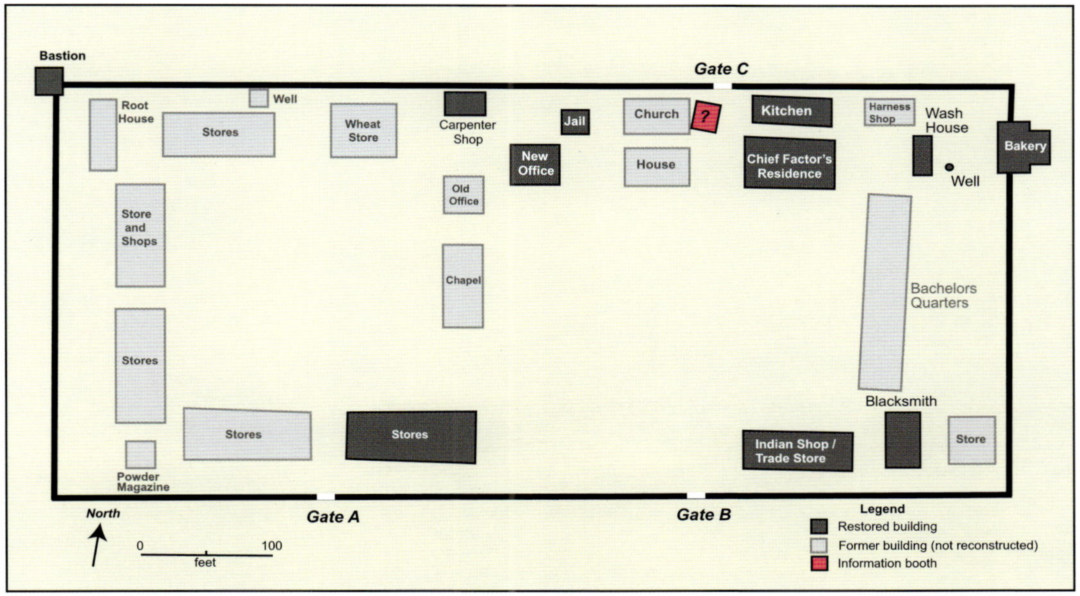

Plan of Fort Vancouver
—Drawing by Philip Mobley

Fort Vancouver today
—Photograph by Paul McDermott

Plate XLIV. *Fort Vancouver, W.T.* Artist: Sohon.

Manuscript drawing of Fort Vancouver by Gustavus Sohon —Courtesy of Yale University Art Gallery and used by permission

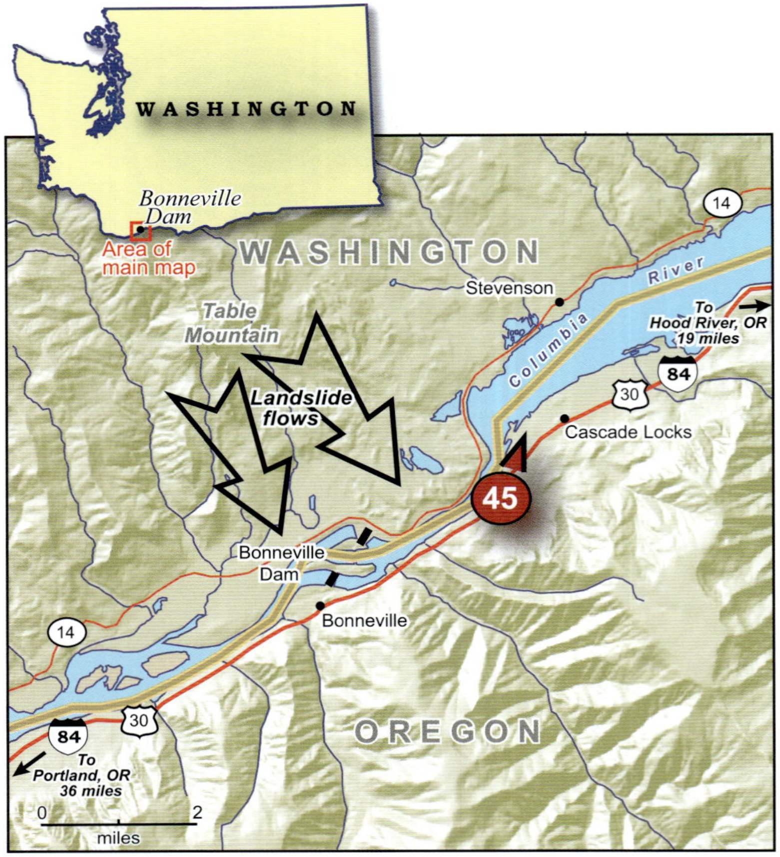

On the evening of November 14, before reaching Fort Vancouver, Stevens camped near the Cascades of the Columbia in his treacherous journey down the Columbia River. His description of this location was very brief:

> I reached the Cascades where I passed the night. Here I met several gentlemen—men who had crossed the plains, and who had made farms in several States and in Oregon or Washington—who had carefully examined the Yakima country for new locations, and who impressed me with the importance of it as an agricultural and grazing country.[104]

Interestingly, Stevens did not comment more about the Cascades and the difficulty they posed to river travel—these rapids were a major obstacle to traffic and over the years took a severe toll in the lives and property of river travelers. In all probability, Stevens and his small party avoided confronting the Cascades directly by portaging.

The Cascades originated from huge landslides rolling off the rock walls and mountains that lined the Columbia River Gorge on its north side. The slides principally originated from Table Mountain, a 3,500-foot-high peak several miles north of the river. Upon reaching the valley floor, the landslides created large earthen dams that temporally curtailed the Columbia's flow downstream. As a result, temporary lakes formed upstream, inundating trees along the banks. When the lake waters became sufficiently deep to breach the dams, the flow resumed, but huge boulders, downed trees, and other debris that had been dumped into the channel remained, creating turbulent waters downstream.[105] When Lewis and Clark traveled along the river in 1805–6, they referred to the narrower channel, which is depicted in Stanley's illustration as "the Great Shute."[106]

Stanley captured the scene nicely. He viewed the Cascades from the Columbia's north shore, looking upstream toward the northeast. In his image, huge boulders protrude above the water's edge—evidence of the historic landslide. Notice how narrow the Columbia is in the "shute," while upstream the river is considerably wider and calmer.

In the accompanying 1889 photograph, a steamboat is beginning a successful run of the turbulent waters. It was a momentous event, as attested by the crowds lining the riverbanks.

Huge, fan-shaped deposits of earth and rock from the landslides still define the river course, but like many other landmarks on the Columbia, the river's once turbulent channel has disappeared. It has been submerged by the reservoir behind the Bonneville Dam, built in 1938. The dam made it possible for watercraft of many sizes to travel up and down the river safely. Before the dam, roads and railroads on the shore provided the only safe passage.

The Cascades in 1929, before construction of Bonneville Dam —Courtesy of Corps of Engineers

A steamboat runs the Cascades, 1889 —Courtesy of University of Oregon and used by permission

Plate XLV. *Cascades of the Columbia.* Artist: Stanley.

Manuscript watercolor of the Cascades by John Mix Stanley —Courtesy of Yale University Art Gallery and used by permission

ON NOVEMBER 15, AS STEVENS WAS HURRIEDLY moving down the Columbia, he passed by Cape Horn. At this point, he was still twenty-five miles from Fort Vancouver. The governor provided no commentary whatever about the site, but a sketch was included in the report, suggesting that Cape Horn had local importance.

Cape Horn is a huge basaltic outcrop on the Washington shoreline of the Columbia River Gorge. The cliffs form a major landmark, in part because they mark the western entrance to the gorge, but also because of the hearty winds that blow there. The place was so named for Cape Horn at the tip of South America; when ships rounded the cape, they encountered fierce winds. On the Columbia River, winds sweeping through the gorge create rugged waves. Today, windsurfers take advantage of the strong winds to maneuver their craft on the river.

Like many other artists and photographers, Stanley found that dramatic cliffs made for captivating scenes. His image is peaceful, indicating that the scene was rendered in spring or summer, not in November when the weather in the Pacific Northwest tends to be stormy. Notice how luxuriant the foliage is—not a sign of late autumn. The two men resting in the foreground do not suggest a survey party in a hurry to get to Fort Vancouver. Unlike most of the other images used in the report, this one was most likely created when Stanley visited the Columbia Plateau in 1847.

Cape Horn, Washington —Photograph by Ronald Grim

Columbia River Gorge as seen from Cape Horn
—Photograph by Ronald Grim

Plate XLVI. *Cape Horn—Columbia River.* Artist: Stanley.

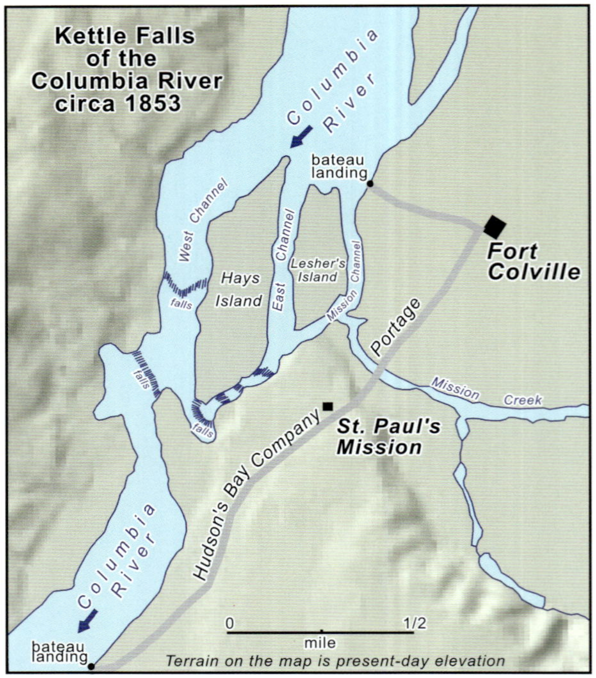

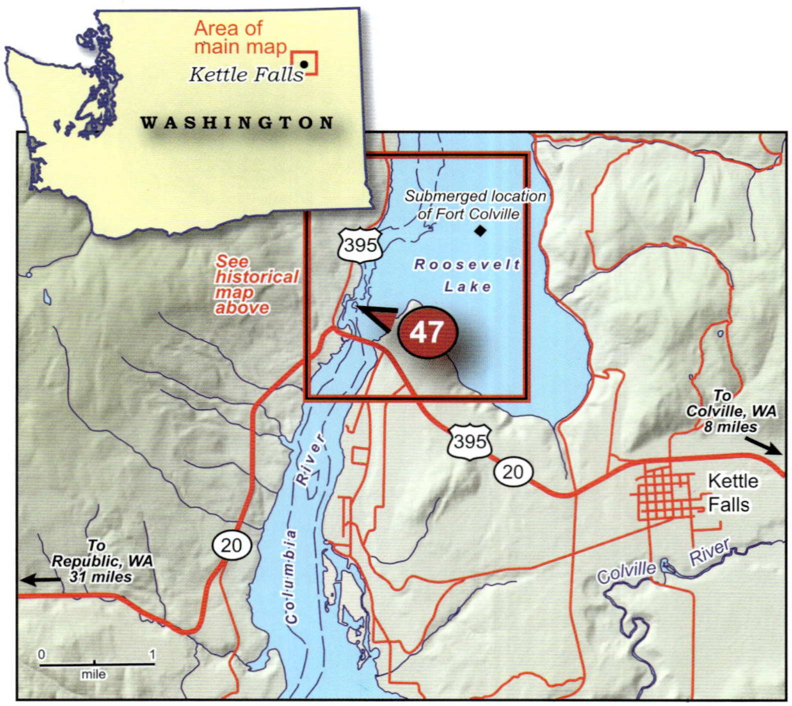

Although Stevens and McClellan rendezvoused at Fort Colville from October 18 to 20, 1853, Stevens made no mention of nearby Kettle Falls in his daily narrative for that time period. Rather, he referenced the falls and an accompanying sketch when he reported on the side explorations of Lt. Richard Arnold of the Third Artillery. Stevens had directed Arnold to ascend the Columbia River from Fort Colville to the 49th parallel in order to explore and determine the latitude of the mouth of the Pend Oreille River (then known as Clark's Fork).

> The Columbia at Fort Colville is about three hundred and fifty yards wide just above the Sometkmuu, or Kettle Falls. These consist of two pitches, one of fifteen feet and another below it of ten, and the river is narrowed to two hundred yards. (See sketch.) An extensive and fertile bottom-land borders the river here, and extends back towards the east for some miles. This continues to border the river, with an average width of half a mile, for twenty miles down the stream, when the wooded hills come closer to the banks, higher and with numerous ravines opening on the river, which makes a route close to its banks impossible.[107]

Stanley's image shows the wild beauty of the falls and documents natives fishing in its turbulent waters. Native fishing here focused on migrating fish, principally salmon as they moved upstream toward their spawning sites. Yet this was a very difficult and treacherous place to fish, as attested by the falls and turbulent water. In spite of the challenges, various native groups harvested an ample supply for their yearly needs. They used different fishing techniques and tools, including spears and traps. The building of the Grand Coulee Dam terminated this activity, because it blocked anadromous fish from moving upstream. The falls themselves were engulfed by the dam's reservoir, Roosevelt Lake.

Fur trappers called these falls *les Chaudieres* (the Kettles). Those who observed the falls probably noticed the large holes carved into the bedrock. These cavities were made by the grinding action of rocks (grinders) twirled around by turbulent water. Geologically, these forms are known as "potholes." In the nineteenth century, the holes were reportedly 15 to 20 feet deep and about 8 to 10 feet in diameter. To the trappers, these huge cavities resembled kettles. Natives referred to the place by several names, including *Ilth koy ape,* meaning "kettle trap-net," and *Shwan-ate-koo,* meaning "deep-sounding waters." [108] All of these place-names reflect either a shape or sound associated with water movement.

Fort Colville, a Hudson's Bay Company trading post, was located on the Columbia's eastern shore within two miles of the falls. Close to the post, the company also erected a small gristmill and a sawmill (see plate 38).

Early-twentieth-century image of Kettle Falls
—Courtesy of Don Popejoy and Spokane Public Library

Kettle Falls region today
—Photograph by Paul McDermott

Plate XLVII. *Kettle Falls, Columbia River.* Artist: Stanley.

Manuscript watercolor of Kettle Falls by John Mix Stanley —Courtesy of Yale University Art Gallery and used by permission

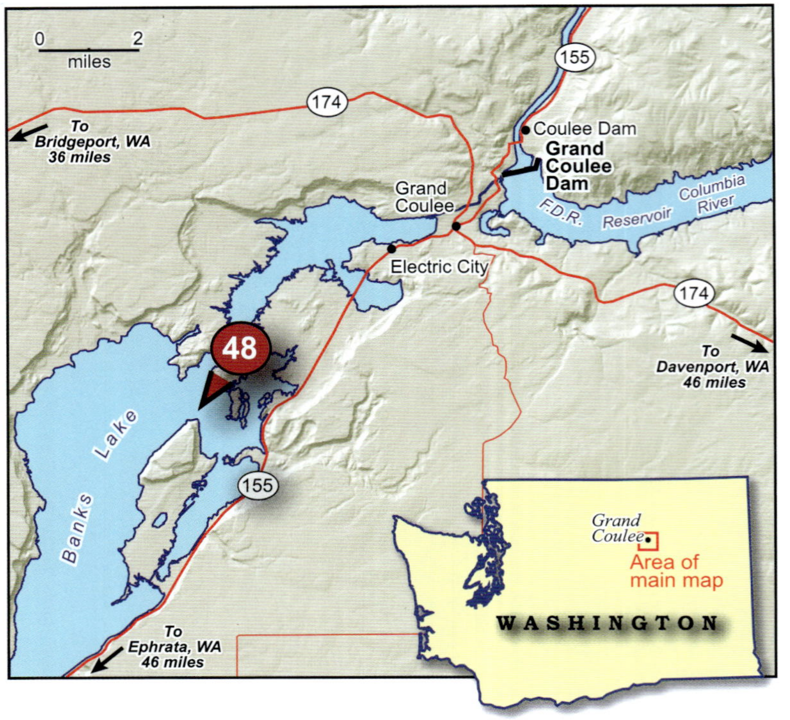

The Grand Coulee near Ship Rock —Photograph by Charles Baumgaertner

After exploring the Columbia River north of Fort Colville, Lieutenant Arnold returned to the post, traveling southwestward across the Columbia River Plateau. Near the current site of Grand Coulee Dam, in Washington state, he encountered the Grand Coulee, a steep canyon etched into basalt. Stevens used Arnold's observations to describe this feature:

> The river runs far below the general level of the Great Plain. The Grand Coulée (see sketch) is about ten miles wide where it opens on the river at its northern end, which is a hundred feet above the water, and gradually widens towards the south; its walls, eight hundred feet high, are formed of solid basaltic rock, but diminish in height southward as the bottom rose towards the summit of the plain, until, in twenty miles distance, they ended. Numerous lateral ravines and cañons were seen running in various directions, some of them containing lakes without outlet, and streams ten feet wide and two feet deep.[109]

The canyon was a product of gigantic floods from the northeast. The floodwaters originated from Glacial Lake Missoula, created behind a continental glacier that formed an enormous ice dam in what is now northeastern Washington and northern Idaho (see plate 32).

When the glacial dam gave way, 500 cubic miles of pent-up water was suddenly released. The floodwaters roared across the landscape at speeds approaching sixty-five miles per hour, draining the lake in as little as fifty-two hours. Over thousands of years, the cycle was repeated again and again. These floods produced considerable erosion downstream—especially on the Columbia River Plateau, in the region known as the Scablands and Palouse Hills.

As seen in this view, the artist was captivated by the virtually vertical walls carved into the basaltic flows. Individual lava flows are easily observed as distinctive layers stacked one atop another. The people in the scene provide scale, though they may not have been present when the image was made. This sketch was most likely done when Lieutenant Arnold explored the region, which means the attribution to Stanley may not be correct since Stanley was not with Arnold. Perhaps someone in Arnold's party sketched the landform, then Stanley re-created it as a watercolor and later as a lithographic print.

The original Grand Coulee was a grandiose structure, but some of its beauty and size have been lost due to its current use as a reservoir for irrigation water. Pictured here is the Ship Rock section of the Grand Coulee, showing the reservoir.

Plate XLVIII. *Grand Coulee.* Artist: Stanley[?]

Big Hole Prairie, looking south —Photograph by Paul McDermott

DURING THE WINTER OF **1853–54**, after Stevens reached Olympia, Washington, his subordinate Lt. John Mullan led a party to thoroughly explore the intermontane region extending from the Canadian border north of Kalispell, Montana, to Fort Hall (near Pocatello, Idaho) on the Oregon Trail. Gustavus Sohon accompanied Mullan on this expedition.

In early winter 1853, Mullan was moving south toward Fort Hall following the St. Mary's (today's Bitterroot) River. While crossing the Bitterroot Mountains, he encountered a vast mountain basin. Stevens recorded Mullan's description of this place:

> [H]e had a fine view of Big Hole prairie, (see sketch,) which is about fifty miles long and fifteen wide. Crossing this on the following day, December 5, its western side was found to be covered with snow some six inches deep, though none appeared on its eastern. It is hemmed in by high mountains on every side except the southeast, where Wisdom river [today's Big Hole River] passes out from it.[110]

The accompanying view looks toward the south-southwest. Sohon climbed up the steep mountainside to the north, which overlooked a small mountain stream at its base. The view is impressive and expansive. A large mountain basin bordered by snow-covered mountains dominates the view. A stream (the Big Hole River) is etched below the basin surface. Its course is delineated by a line of small trees on the right. The basin itself is covered with small plants and sagebrush. Stillness pervades the scene, except for several men on horseback crossing the ridge in the center of the illustration.

Twenty-four years later, this remote, peaceful valley became an important site in the history of the Nez Perces and the western United States. On August 9, 1877, the Nez Perces, under the brilliant leadership of Chiefs Joseph and Looking Glass, resisted the U.S. Army in the Battle of the Big Hole.[111]

Plate XLIX. *Big Hole Prairie, from the North.* Artist: Stanley after Sohon.

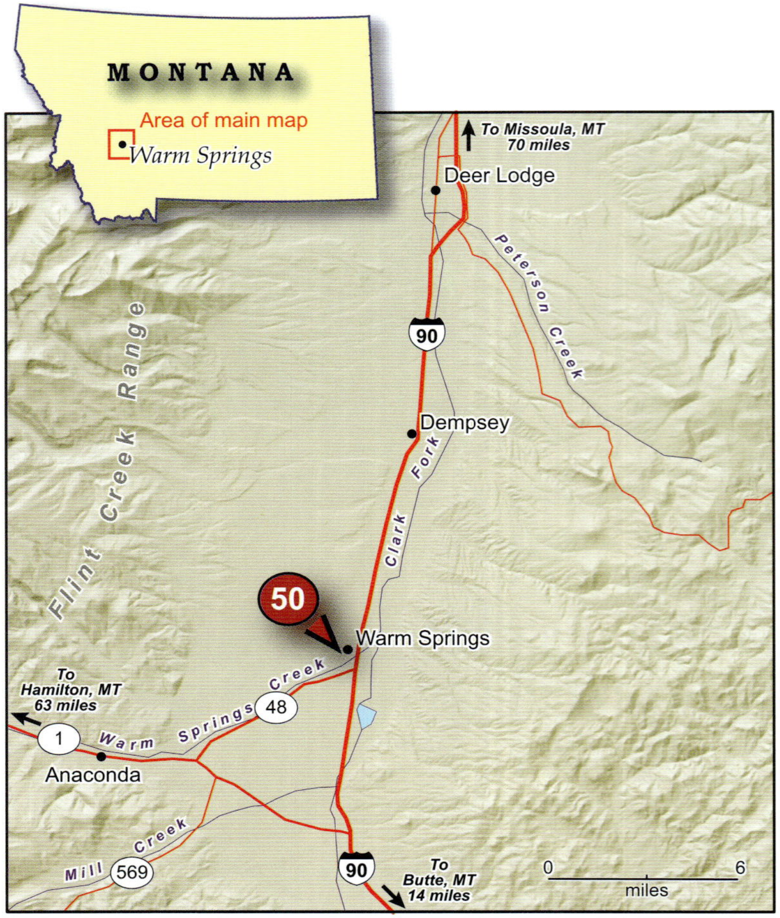

WHILE IN THE DEER LODGE VALLEY IN JANUARY 1854, Mullan and Sohon visited this travertine mound from which hot mineral-laden water flowed. After thousands of years, the mineral deposits accumulated to form a forty-foot-high mound with a thirty-foot diameter. Water issuing from the hole is extremely hot—172°F.[112] Stevens described Mullan's visit to this locale:

> [He] entered the long, level, and beautiful prairie called the Deer Lodge, from the great numbers of deer found about it. It resembles that of the Bitter Root valley, but is more extensive, being about fifty miles long and twelve to fifteen wide, bounded on all sides, except the southeast, by high wooded mountains, of which the summits only were covered with snow. Like that, too, it is noted for the slight fall of snow in it during the severest winters, and its many streams are lined with timber. Large numbers of antelope were feeding in it, and mountain sheep and goats were seen on the peaks around. Near its centre is a conical mound (see sketch) about thirty feet high, around the base of which are innumerable springs of hot water, and on top a spring three feet in diameter, down which was thrust a pole twenty feet long without finding bottom. The water from this does not run over the top. A white soft salt is deposited around it.[113]

In the illustration, we see two men standing at the mound's summit. One of them is attempting to measure its depth using a twenty-foot pole. Sohon's images were sometimes different in style than Stanley's. One difference is demonstrated here—the use of an oval format. None of Stanley's images in the report were oval.

In its pristine form, the mound and its warmer microclimate attracted many animals, including deer. The springs and its surrounding vegetation would be particularly attractive in the winter, when snow covered the valley elsewhere and temperatures were below freezing.

Like many thermal springs, the waters of Hot Spring Mound were thought to have medicinal and therapeutic properties. It is not surprising that a health facility arose here. Hot Spring (or Warm Springs) Mound now sits on the grounds of a state mental hospital, established in 1912. Sadly, the mound's original rust-colored coating has largely disappeared. At one time, a walkway eased the climb up the mound's steep slope, but it is no longer in use because authorities are afraid of someone getting hurt. A large gazebo still sits atop the mound, but the spring has been capped off.

Hot Spring Mound, Warm Springs, Montana —Photograph by Paul McDermott

Plate L. *Hot Spring Mound in the Deer Lodge Prairie of the Rocky Mountains.* Artist: Sohon.

Manuscript drawing of Hot Spring Mound by Gustavus Sohon —Courtesy of Yale University Art Gallery and used by permission

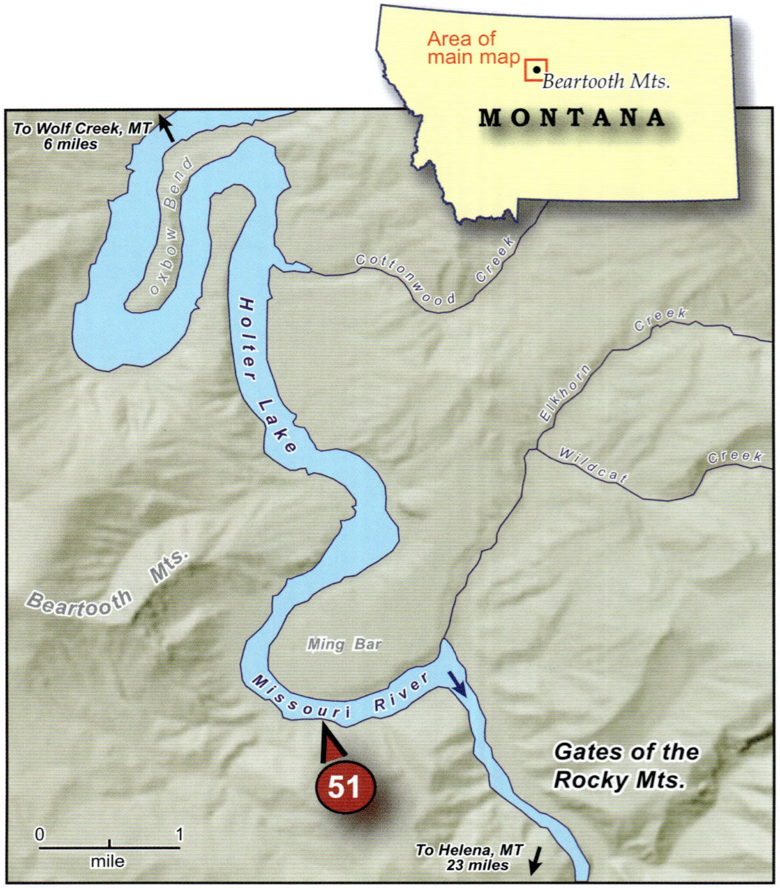

On March 4, 1854, Lt. John Mullan left Cantonment Stevens, his winter base camp in the Bitterroot Valley, and proceeded generally east through the Rockies toward the Missouri River; there he would turn north to Fort Benton. His goal was to test the feasibility of crossing the Continental Divide with a wagon. For a time he followed the Missouri River as it passed through the "Gate of the Mountains," as identified by Lewis and Clark. One of the major landmarks in this vicinity is a jagged mountain outcropping referred to as the "Bear's Teeth," so named for its sharp points, which supposedly resemble grizzly teeth.

This image depicts a singular mountain as its main subject, sloping up from the banks of the Missouri River. The artist shows the mountain in the morning hours, looking toward the north. The accompanying map shows the location of the mountain (now known as Beartooth Mountain) and the artist's probable vantage point. Although Stanley made the watercolor upon which the print was based, whether or not he ever visited this site is open to question. The watercolor may have been based on a sketch that Sohon created in the winter of 1854.

Beartooth Mountain today
—Photograph by Paul McDermott

Plate LI. *Bear's Teeth, Missouri River—Gate of the Mountains.* Artist: Stanley.

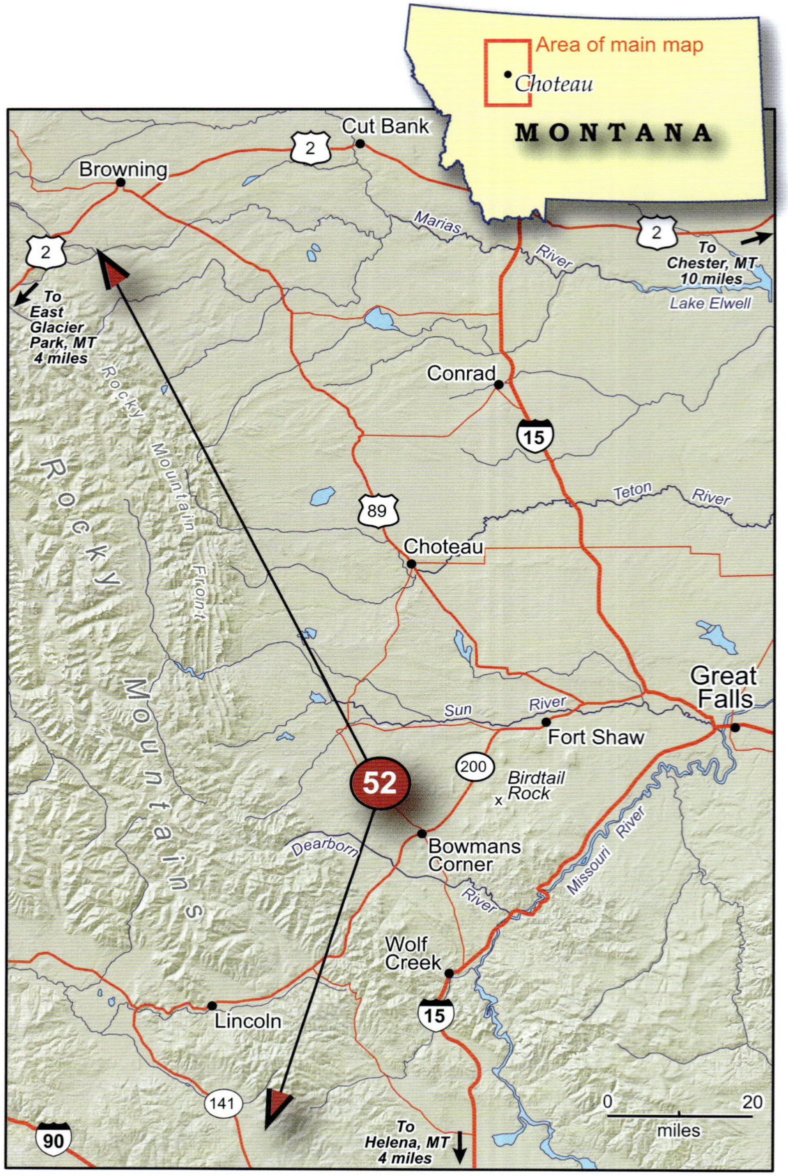

Sohon, the artist of this panoramic view, first saw the main range of the Rockies in the early months of 1854 while accompanying Mullan's party from Cantonment Stevens to Fort Benton. Sohon made this image in March.

This wide-ranging perspective reflects Sohon's skill in drawing wide-angle views. His vantage point is most likely imaginary, as it is impossible to see the entire eighty-mile chain, from Marias Pass to the Dearborn River, from a single point. The image shows drama and action. An Indian hunter lies prone on the ground, in the act of shooting animals in the distance. Mullan's exploration party is seen moving westward toward the mountains on the left horizon.

The Rocky Mountains were created relatively recently in geological history—about 60 million years ago. It took over 30 million years to create the mountain chain as we know it today. Glaciation and stream erosion helped shape the landforms originally created through the mountain-building process called orogeny—a folding, faulting, and uplift of the earth's crust.[114]

Part of the Rocky Mountain Front, near Choteau, Montana
—Photograph by Paul McDermott

Plate LII. *Main Chain of the Rocky Mountains, as Seen from the East, Extending from a Point North of Marias Pass to Near the Little Blackfoot Pass.* Artist: Stanley after Sohon.

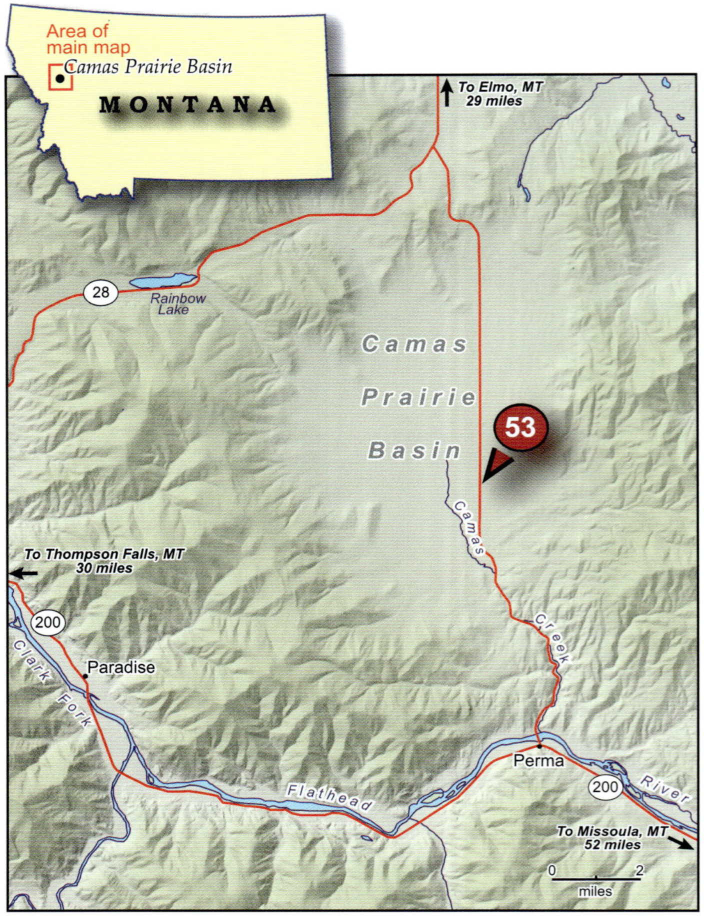

STEVENS REPORTED ON MULLAN'S CONTINUED EXPLORATION during the spring of 1854:

> The following April Lieutenant Mullan also explored the Flathead river and the country north of it as far as the Koutenay, at the 49th parallel. Leaving the Bitter Root valley on the 14th, he followed the route already used by the main expedition, down Clark's Fork as far as Kamas Prairie, (a sketch of which is here given,) twenty miles below the mouth of the Jocko.[115]

This "Kamas Prairie" was a broad basin bordered by small ridges. Here natives living in the region harvested a very nutritious plant—the camas (*Camissia quamish*). The plant resembles a large scallion, with thin green stalks and a small bulb at the root. When baked, insulin in the plant is converted into fructose. Natives harvested the plant using a shovel fashioned from a deer or elk horn.[116]

Earlier in his report (October 8, 1853), Stevens mentioned eating the camas bulb:

> We gave all the Indians coffee, and the women in return gave us cooked kamas root. It is of a dark color, small, between the pear and onion in shape, and of a sweet, agreeable flavor.[117]

Drawing of camas plant by Gustavus Sohon —Courtesy of the Smithsonian Institution

Blue camas plant in flower —Courtesy of Wayne Phillips and used by permission

Plate LIII. *Kamas Prairie of the Pend d'Oreilles Indians, in the Rocky Mountains, Looking Southward.* Artist: Sohon.

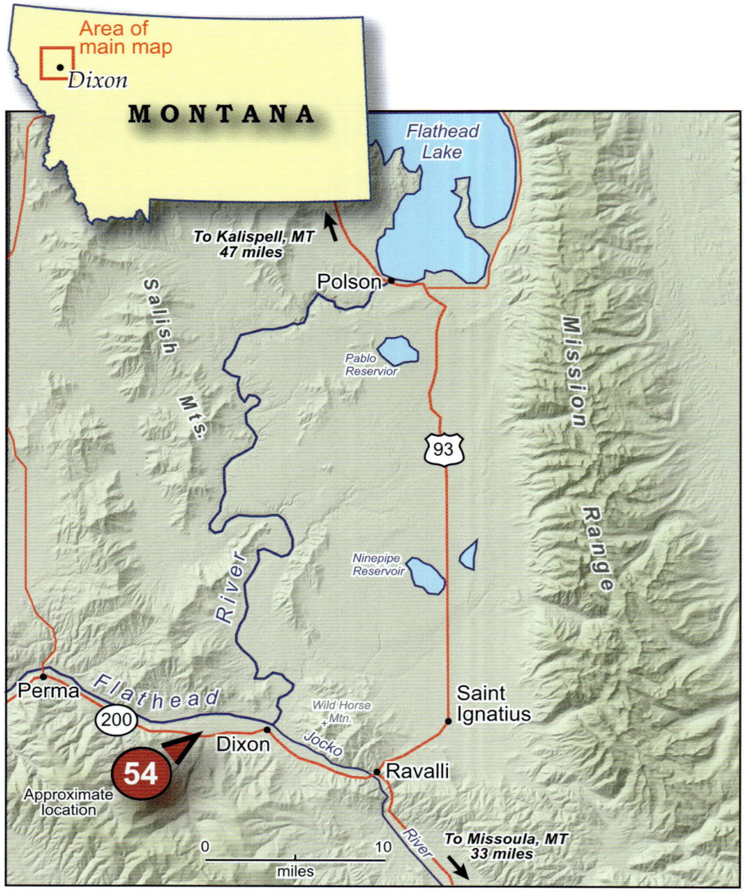

In early April, Lieutenant Mullan and his small party followed the Jocko River into the area around Flathead Lake and north toward the 49th parallel. In five months, they explored this vast intermontane region east of the Bitterroot Mountains. The group focused on obtaining geographical information relating to rivers flowing through the region, especially the Clark Fork, Flathead, and Kootenai. When Sohon rendered this image, this portion of today's Flathead River was identified as Clark's Fork. Stevens reported on Mullan's exploration:

> Spring was then well advanced, and the prairie covered with long grass. Between it and Clark's Fork, towards the lake, he crossed a low ridge of hills, thickly wooded; reaching Clark's Fork again, at the mouth of Hot Spring creek, he traveled up its west bank. The river banks and the adjoining country, being much cut up by coulées, have the appearance of that on the Upper Missouri. (see sketch.) The soil is principally a light-yellow clay; the stream here is two hundred yards wide, swift and deep, sparsely timbered with pine and cedar, and a few cottonwood and aspen.
>
> From a high point he could see over the country for a great distance, limited to the eastward by a high snow-clad range of mountains, which runs close to the eastern border of Flathead lake at its whole length, and continues on northward to the head of Clark's Fork.
>
> Just below the outlet of the lake there is a series of rapids and falls, one of which, at this season, was fifteen feet high. The country to the west of the lake is a high rolling prairie. Salmon trout, three feet long, are caught in it, supplying one of the principal articles of subsistence to the Indians of the country.[118]

Sohon's view portrays the region from a viewpoint south of the Flathead River, looking northeast toward the Mission Mountains, seen in the background. The site is probably near present Dixon, Montana. Notice how coulees punctuate the riverbanks.

Plate LIV. *View of the Clark's Fork and the Ridge of the Mountains, South of the Flathead Lake, Looking Eastward.* Artist: Sohon.

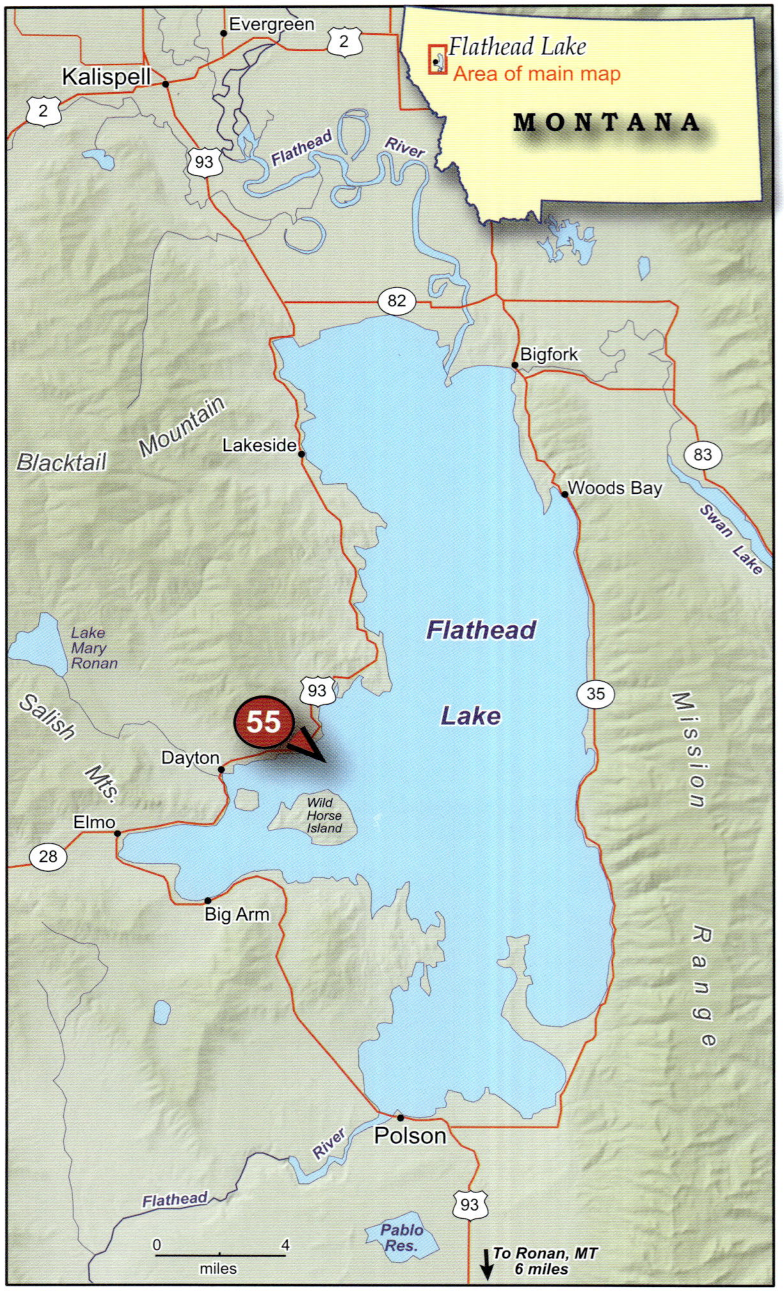

FLATHEAD LAKE IS THE LARGEST FRESHWATER LAKE west of the Mississippi. It occupies approximately 195 square miles and is very deep in its eastern portion—about 370 feet. It owes its origin to glaciation. During the great ice age, large continental glaciers carved and etched their way into the rocks of the Rocky Mountain trench. After a large terminal moraine formed near the lake's southern extension, the glacier eventually retreated through a process known as ablation, a loss of ice due to melting and evaporation. Meltwater collected in the abandoned basin and behind the terminal moraine. The end result was Flathead Lake.

The report recorded Mullan's view of the lake:

> Some distance up the west bank of the lake he had an excellent view of it, (see sketch,) showing that it contained several large and beautiful islands, all covered with an excellent growth of pine. Many of these islands were several miles long. The width of the lake he estimated at eight miles.[119]

Although this image is attributed to Stanley, he never visited the area. It is more likely that the original was drawn by Sohon, who spent several months near and around the lake in 1854. The artist's vantage point was the lake's western shore, probably near present Dayton, Montana, looking to the southeast. The rugged mountains of the Mission Range form a backdrop for the lake, and the low outlines of islands dotting the lake's surface near its eastern shore are easily seen.

Flathead Lake near site where image was drawn
—Photograph by Paul McDermott

Plate LV. *Flathead Lake, Looking Southward.* Artist: Stanley[?]

DRAMA PERMEATES THIS IMAGE. It was early spring, and meltwater filled the Hell Gate (today's Clark Fork) River beyond its banks. Mullan and his party, including Sohon, were returning to Cantonment Stevens from their exploration of northern Montana around Flathead Lake. To reach the camp, they had to cross the Hell Gate River near present-day Missoula. They built a simple log raft for this purpose, but it proved to be inadequate in the rushing current, and the men almost drowned. The raft drifted downstream and got snagged on a tree anchored to the stream floor, known as a sawyer. The men saved themselves by holding on to the tree's extended roots and trunk. Stevens briefly noted the incident:

> [O]n the 5th of May arrived at Cantonment Stevens in the Bitter Root valley, without any difficulties except in crossing the flooded rivers, among which, at the Hell-Gate, [Mullan's] whole party and property were nearly lost in using a raft unmanageable in the swift current. (see sketch.)[120]

Sohon's image, a reconstruction of the actual event, shows the small raft slightly submerged below the river's surface, entwined in the sawyer. Of all the images created for the Stevens report, this is the only one that portrayed people in large scale. Three men dominate the scene. Two of them were probably Mullan and Sohon.

Clark Fork (Hell Gate) River near Missoula, Montana
—Photograph by Paul McDermott

Plate LVI. *Crossing the Hell Gate River, May 5th 1854*. Artist: Sohon.

STEVENS'S PACIFIC RAILROAD EXPEDITION EFFECTIVELY ended in September 1854. In this month, Lt. John Mullan finished his portion of the survey by crossing the Bitterroot Mountains and returning to Fort Vancouver. Stevens reported on Mullan's movement west from Cantonment Stevens and the Bitterroot Valley:

> Leaving the Bitter Root valley again, on the 19th of September, Lieutenant Mullan went down the river to Lou-Lou Fork, which is fifteen yards wide and two feet deep at its mouth. Its valley is five hundred yards wide and the mountains on each side are quite high and well timbered with pine and cedar (*Arbor Vitae*). (see sketch.) A good road is found for twenty miles up the stream, partly through prairie and open forest.[121]

Lewis and Clark used this route, which followed a Native American trail, to cross the Bitterroot Mountains in September 1805. The route has since become known as the Lolo Trail. Lolo Pass, as the eagle flies, is thirty-one miles southwest of today's Missoula. Mullan made a cursory exploration of the route, creating one reconnaissance map and a few images to document the probe.

This view by Sohon shows the Mullan party entering the Bitterroot Mountains near Travelers Rest—a campsite Lewis and Clark used in 1805 and again in 1806. A low line of scrubs, on the image's right side, marks the course of Lolo Creek. Several trails cross the scene; each delineates a major Native American trail going either into the mountains or through the Bitterroot Valley in a north–south direction.

Aerial view of Lolo Creek, Lolo, Montana —Courtesy of Jim Wark, from *Discovering Lewis and Clark from the Air* by Joseph A. Mussulman and Jim Wark; used by permission

Plate LVII. *Entrance to the Bitter Root Mountains, by the Lou Lou Fork.* Artist: Sohon.

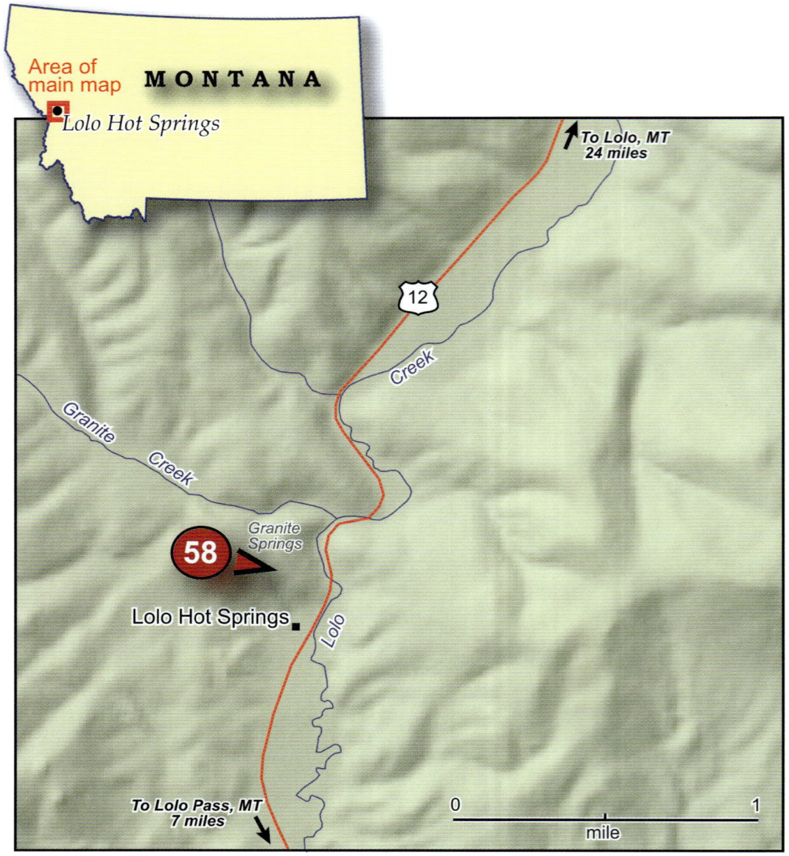

LOLO HOT SPRINGS IS A PLEASANT SITE along U.S. route 12, which connects communities now located along portions of the trail through the Bitterroots. The Nez Perces called this trail *Kuseyne 'Iskit,* the buffalo trail, identifying one of trail's major uses.[122] But it was also used to access berries in and around the Travelers Rest region.

Mullan traveled west along the trail in September 1854. His goal was to determine the route's feasibility for use by a railroad. He was not impressed:

> Above this the route is over rugged hills to the dividing ridge, which was low and obstructed with fallen timber. . . . The route is thoroughly and utterly impracticable for a railroad.[123]

Like the Lewis and Clark expedition, Mullan's small party found a pleasant set of hot springs just east of the range crest:

> Near the summit of the mountain [Mullan] found a hot spring, with a temperature of 132°, around which was a fine prairie camping ground. (see sketch.) . . . This route is nearly one and the same as that followed by Lewis and Clark in 1805; the Lou-Lou Fork being their Travellers' Rest creek, and the hot springs referred to are those spoken of by these indefatigable explorers.[124]

Plate LVIII. *Hot Springs at Source of Lou Lou Fork, Bitter Root Mountains, Looking West.* Artist: Stanley[?]

Rock formation near Lolo Hot Springs —Photograph by Paul McDermott

Lewis and Clark provided more precise detail for the springs:

> These warm springs are situated at the foot of a hill on the north side of Traveller's-rest Creek, which is ten yards wide at this place. They issue from the bottoms, and through the interstices of a gray freestone rock, which rises in irregular masses round their lower side. The principal spring, which the Indians have formed into a bath by stopping the run with stone and pebbles, is about the same temperature as the warmest bath used at the hot springs in Virginia [Bath, Virginia, now Berkeley Springs, West Virginia].[125]

Commercial development of these hot springs began in 1883.[126] Through the years, its commercial value has waxed and waned. Currently, the site contains a swimming pool with lodging and eating establishments.

The region is filled with granite columns that have been subjected to spheroidal weathering. In the lithographic print, one such column dominates the left edge of the view. While today the landscape is fairly well wooded, the artist depicted a more barren scene. Our photograph, taken in May 2006, shows a picturesque view filled with trees, marshland, and a mountain stream gushing through the small valley. A question arises as to the disparity between the two images created 152 years apart. Possibly, the climate was colder in 1853 than it is now, so trees were less abundant. Or perhaps the current vista reflects the impact of recent forest fires.

This image was attributed to Stanley, but this was a mistaken citation. It was the work of Sohon, who accompanied Mullan through the Lolo region in the late summer of 1854. Stanley was not there, having returned to the East Coast late in 1853.

Lolo Creek today —Photograph by Paul McDermott

THIS SMALL CAMP SERVED AS THE BASE for John Mullan's explorations of the northern Rockies during the winter of 1853–54. Mullan was directed by Stevens to collect meteorological data defining winter conditions prevailing over the entire region. In particular, Stevens was very concerned about winter snow, because it would affect both railroad construction and rail operations.

Originally, Fort Owen was to be Mullan's headquarters for this assignment, but he was not satisfied with the fort and subsequently moved his base of operations on October 8, 1853, ten miles south to a site in the Bitterroot Valley. He named the camp Cantonment Stevens. Stevens explained Mullan's reasons for the move:

> [Mullan] established his camp ten miles above Fort Owen, at a point where there was excellent grass, wood, and water, and where, in consequence of its being a little removed from the Indian camps, he could better regulate the intercourse of his men with them.[127]

As Sohon showed in this image, four cabins formed the camp's core. One cabin was for the officers, the others for enlisted men and civilians. The complex was completely enclosed, and in its center was a flagpole as tall as the surrounding trees. Cantonment Stevens appears to have been a well-planned post, situated next to good grazing and ample wood. Although Stevens remarked that Mullan built the post "a little removed from the Indian camps," the image shows seventeen tepees (about 170 people) in close proximity to his headquarters.

Another intriguing feature found in both the watercolor and lithographic versions of this view is the depiction of two Indians positioned near the central focal point. Each holds a bow and arrow, and they are apparently taking aim at a target on the opposite side of a ravine. However, this feature was deleted when Mullan included a similar image, also attributed to Sohon, in his 1863 *Report on the Construction of a Military Road from Fort Walla-Walla to Fort Benton*. Was this a conscious editorial decision? If so, was Mullan attempting to convey a subtle message that the presence of the Native American population would not pose a threat to the construction of the military road?

Early maps such as the one illustrated here, documenting John Mullan's surveys of 1853–54, indicate that Cantonment Stevens was located ten to twelve miles south of Fort Owen. That would put site near today's Corvallis, Montana, along the banks of Willow Creek, a tributary of the Bitterroot River. However, no one yet has found archeological evidence demonstrating the camp's precise location.

Detail of Bitterroot Valley showing the approximate location of Cantonment Stevens, from John Mullan's manuscript map in Sketch of a Reconnaissance across the Bitter Root Mountains to Fort Dalles, Oregon *(1854)* —Courtesy of the National Archives and Records Administration

Lithographic view of Cantonment Stevens by Gustavus Sohon, published in Mullan's Report on the Construction of a Military Road from Fort Walla-Walla to Fort Benton *(1863)* —Courtesy of Geography and Map Division, Library of Congress

Plate LIX. *Cantonment Stevens, Looking Westward.* Artist: Sohon.

Manuscript drawing of Cantonment Stevens by Gustavus Sohon —Courtesy of Yale University Art Gallery and used by permission

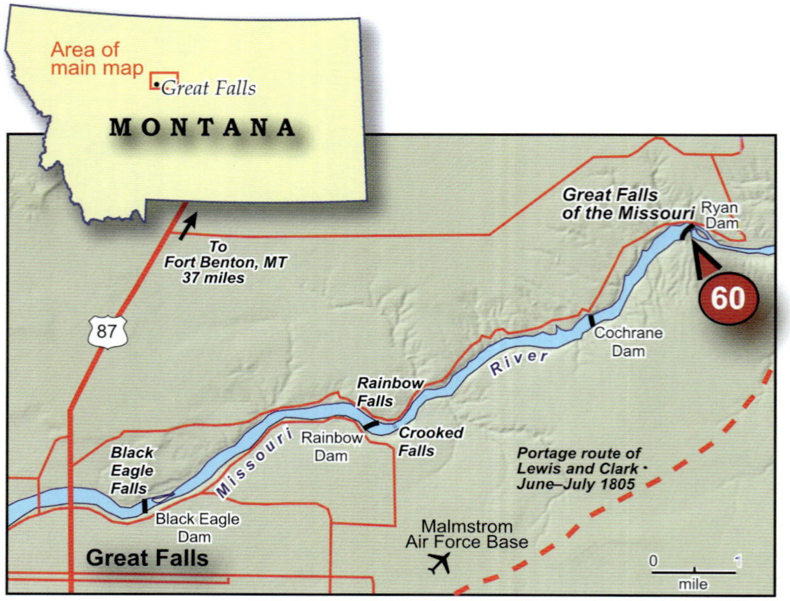

Lithographic view of the Great Falls by Gustavus Sohon, published in Mullan's Report on the Construction of a Military Road from Fort Walla-Walla to Fort Benton *(1863)* —Courtesy of Geography and Map Division, Library of Congress

Between Great Falls and Fort Benton, Montana, the Missouri River descends into a deep canyon. In the process, it drops through a series of five major falls and numerous rapids, which proved to be both fascinating and frustrating to explorers, including Lewis and Clark. The large falls blocked travel up and down the river, yet their thundering sound and swirling turbulent waters were captivating. Artists like Stanley and Sohon spent many hours sketching and painting these grandiose cascades.

Several members of the Stevens party visited the Great Falls area, including Lt. Cuvier Grover, who was sent to explore the Missouri River late in 1853. He described the falls region in his report:

> The Great Falls of the Missouri are situated about 75 miles by land above the fort [Fort Benton]. There are five principal cascades; the first of twenty-five feet fall, about three miles below the mouth of Sun river; the second nearly three miles below this, of five feet eleven inches; and immediately below it the third. Here, between high banks, a ledge, nearly as straight as if formed by art, runs obliquely across the river and over it the waters fall forty-two feet in one continuous sheet of four hundred and seventy yards in width. Half a mile below this is the fourth, a small, irregular cascade of about twelve feet descent. The stream then hurries on, lashed and churned by numerous rapids, about five miles further, when it precipitates itself over a precipice seventy-six feet high. The banks are high and abrupt on both sides, with deep ravines extending into the prairie for one or two miles both above and below it. Below the falls there is a continuation of rapids, which become less and less frequent, to the mouth of Highwood creek. Steep banks about two hundred feet high crowd closely upon the river, and on the north side are so cut up by precipitous ravines that it is almost impossible to keep near it.[128]

Sohon's illustration, which was framed within an oval, featured the seventy-five-foot precipice. The view was toward the west-southwest; he was positioned about a mile downstream from the falls. Water vapor rises above the cascade on the left side. Like many of Sohon's original drawings, the manuscript version of this image was created in pencil on three-toned paper. The clouds were fashioned by removing paper pigment using a sharp, pointed instrument.

Sohon made four different images of the Great Falls. Several of these were created while the artist was under the leadership of John Mullan in 1859–62. One of these, which was published in Mullan's 1863 report on the construction of the military road, is similar to the one contained in the Stevens report, but its scale is larger, and the vantage point is closer to the water.

Today, only a portion of the Great Falls remains. The construction of Ryan Dam obliterated most of the natural river system.

Upper Falls, one of five major falls at Great Falls
—Photograph by Paul McDermott

Ryan Dam at Great Falls
—Photograph by Paul McDermott

Plate LX. *Great Falls of the Missouri River.* Artist: Sohon.

Manuscript drawing of the Great Falls of the Missouri River, by Gustavus Sohon —Courtesy of Yale University Art Gallery and used by permission

Stevens first visited the Palouse River Valley with his survey party in the fall of 1853 (see plates 40 and 41), then again when he was conducting treaty negotiations in 1855. On the former trip he reported on its suitability as a possible route for a railroad or a military wagon road:

> Before descending into the valley of the Peluse, I ascended a very high hill on the last divide, and westward of our trail, where I had a most distinct and interesting view of the sources of the Peluse. (see sketch.) So much was I impressed with it, that I directed Mr. Doty to carefully take notes and lay down the river on his map. It has its source in the main ridge of the Bitter Root, sixty or eighty miles distance from the hill where we observed it, and flows in nearly a straight course through a valley some twenty miles wide, bearing north 77° east, through a country densely timbered with pine.[129]

The older spelling of this river's name is *Peluse,* while the modern spelling is *Palouse*. The river, which is nearly 140 miles long, officially begins in Latah County, Idaho, then gradually moves west-southwest toward its juncture with the Snake River in Franklin County, Washington. Approximately six miles above its junction with the Snake, the Palouse drops over a beautiful 198-foot waterfall (Palouse Falls), then moves through a precipitous canyon. Prior to dam building on the Snake, there was a deep hole in the Snake River near its junction with the Palouse; the hole was a primary fishing spot for the Palouse peoples who periodically camped nearby.

Sohon was the first explorer to map in detail the course of the Palouse. He did this work in 1859. Most of this exploration was done in conjunction with the construction of John Mullan's military wagon road from Fort Walla Walla to Fort Benton. Some of Sohon's notes and early maps relating to the Palouse exploration are now preserved in the vaults of the Geography and Map Division of the Library of Congress in Washington, D.C. Ultimately, the valley did not provide good access for a passage through the Bitterroot Mountains, and the route was subsequently abandoned.

Sohon's image shows the broad valley where the river originates. The valley is dominated by patches of coniferous evergreen forest. In between, grassland covers the valley floor. The outline of mountain ridges in the background are either the foothills or the main ridge of the Bitterroots. In the foreground are picturesque shrubs scattered over a small hill.

Plate LXI. *Source of the Peluse.* Artist: Stanley after Sohon.

STEVENS EXPLORED THE BIG BLACKFOOT VALLEY extensively in September 1853 and again in 1855. Stevens recorded his impression in the report:

> In order to get a view of the country we immediately ascended some very steep bluffs to the eastward, towards the divide of the mountains, and finally reached a point where the whole country lay before us. Here we saw the rolling prairie over which we had passed this morning, the stream from Lewis and Clark's Pass winding on its left through the rocky cañon, and at our feet a very low divide, which made it almost a matter of demonstration to us that we could pass from the valley stream by a low cut to the open prairie, and gain the Big Blackfoot by the stream up which we had come. Immediately pushing down to the low divide referred to, I caused a sketch of the country to be made. . . . Starting now from the point where the trail from Lander's Fork strikes the stream flowing from Lewis and Clark's Pass, we continued up the latter stream for five miles, passing over the most remarkable valley that I have ever seen in the immediate region of a mountain divide (see sketch).[130]

The (Big) Blackfoot River begins near the Continental Divide and flows westerly 132 miles to its junction with the Clark Fork, near present Bonner, Montana. For years, the Blackfoot River valley functioned as a major Indian trail through the Rocky Mountains. Stevens and later John Mullan thought the Blackfoot valley would provide a viable railroad route. Consequently, while working on his military road project during the dreadfully cold winter of 1861–62, Mullan established a construction camp, known as Cantonment Wright, on the south bank of the Blackfoot River near its juncture with the Clark Fork. Today, Montana state route 200 traces the explorers' route.

Glacial ice and flooding played a role in forming the Blackfoot's valley. The hummocky surface shown in the Sohon image reflects glacial erosion and deposition. Several small glacial boulders (erratics) punctuate the foreground, while prominent coniferous trees are distributed throughout the image. The exact location represented by this print is not known. We surmise that it portrays the Big Blackfoot prairie, also known as Prairie of the Knobs.

Lithographic view of Cantonment Wright at the juncture of the Blackfoot and Clark Fork Rivers, by Gustavus Sohon —Courtesy of Special Collections and University Archives, University of Oregon Libraries

Plate LXII. *Big Blackfoot Valley.* Artist: Sohon.

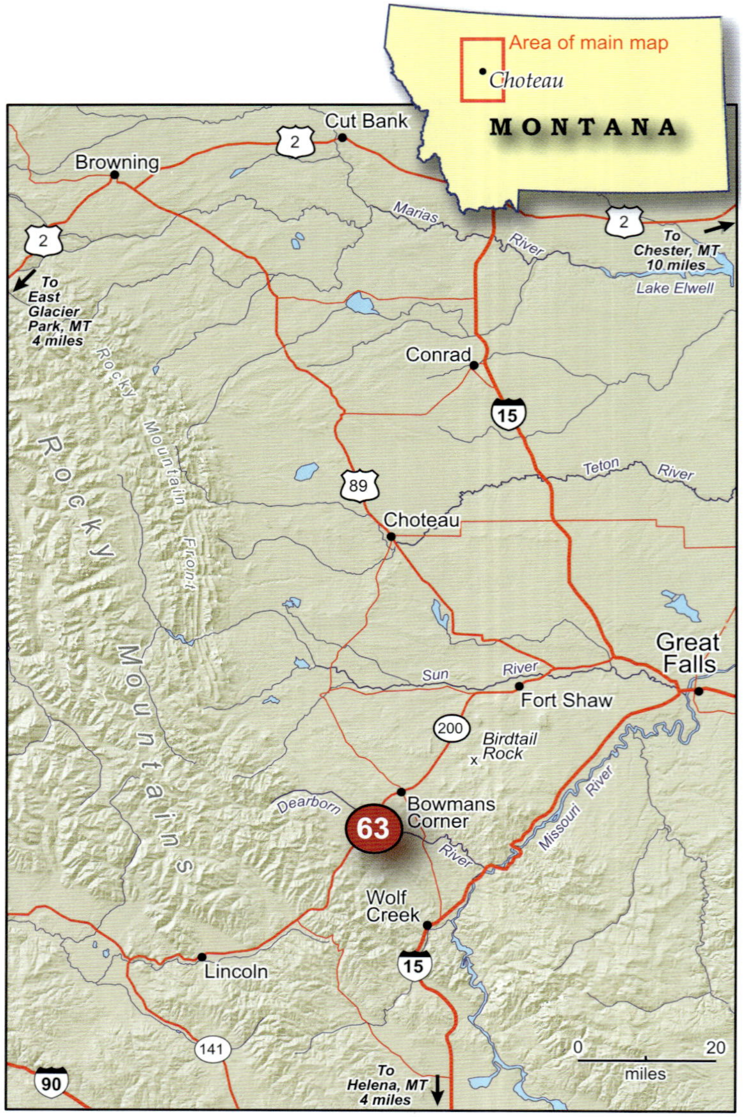

IN A GEOGRAPHICAL SUMMARY, Stevens described the region on the east side of the Rocky Mountains:

> Through this region are many landmarks which will attract the especial attention of the voyaguer. Besides these [mountains] there are prominent buttes near the line of travel, often of striking and fantastic forms, such as the Boque d'Otard, Knee, and Trunk, between the Marias and Teton; the Oksut, south of the Teton; the Big Knees, Crown Butte, and Birdtail rock, near Sun river; and Heart mountain, (of which a sketch is given), near the sources of Elk Fork. . . . The plateau itself is arable in many portions, though not often continuously so.[131]

Following U.S. route 89 west from Sun River takes motorists to Choteau, Montana. As the road moves toward the northwest, one can see the landscape depicted in this view. The sketch illustrates the general region, its expanse defined by two letters. The letter *a* on the left designates the location of Lewis and Clark Pass; the *b* on the right identifies Heart Mountain. The image condenses the distance between the two points. In the foreground, a small pack train moves over a rolling landscape. The inclusion of the pack train suggests that Sohon was the artist, even though he is not identified as such—Sohon often included mules in his images. Mules also appear in a Sohon drawing of Birdtail Rock, shown here. We know Sohon visited this locale in 1854, on his trip with John Mullan from Cantonment Stevens to Fort Benton.

Drawing of mules passing Birdtail Rock, by Gustavus Sohon, 1854 —Courtesy of National Archives and Records Administration

Plate LXIII. *Rocky Mountains, Looking Westward.* Artist: Not identified.

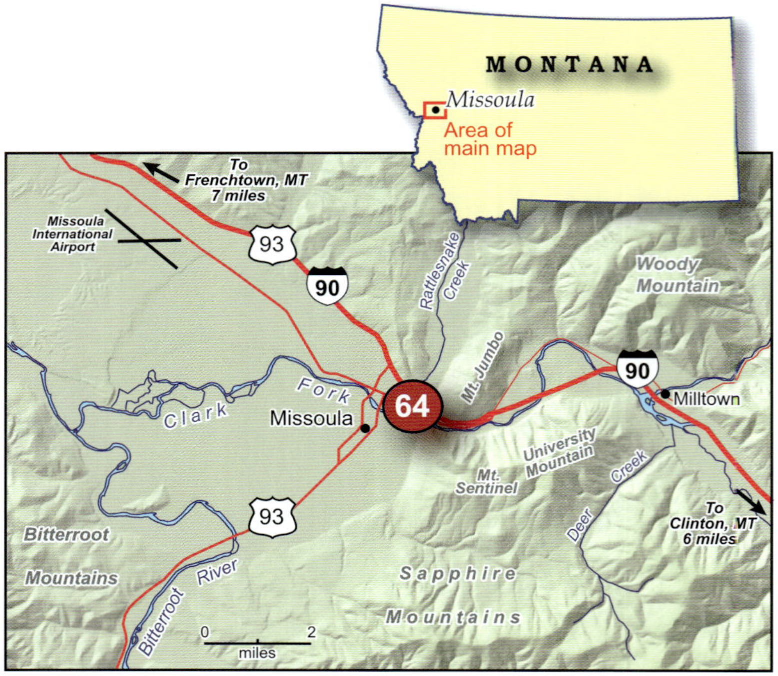

Modern view of the junction of the Blackfoot and Clark Fork Rivers, about five miles upstream from the entrance to Hellgate Canyon depicted in plate 64 —Photograph by Paul McDermott

At Missoula, Montana, the Clark Fork River enters a large valley formed by ancient Lake Missoula. The Clark Fork's entrance into the Hell Gate Ronde is prominently marked by two large mountains, Mounts Jumbo and Sentinel, which are illustrated in Stanley's lithograph. Stevens, in his geographical summary, described the complex network of rivers that converge at this location:

> From its lower end, at the junction of the Hell-Gate [Clark Fork east of Missoula], it is believed that the Bitter Root [today's Bitterroot, also known as St. Mary's] river is, or can be made navigable for small steamers for long distances, at least, thus affording an easy outlet to its products in a natural direction. Hell-Gate (see sketch) is the debouche of all the considerable streams which flow into the Bitter Root [Clark Fork west of Missoula], eighty-five miles below its [today's Bitterroot's] source at the Big Hole divide. The distance from Hell-Gate to its junction with the Bitter Root is fifteen miles. It must not be understood from the term Hell-Gate that there is a narrow passage with perpendicular bluffs; on the contrary, it is a wide, open, and easy pass, in no case being less than half a mile wide, and the banks not subject to overflow at all. At Hell Gate is the junction of two streams; the one being the Hell Gate [Clark Fork] river and the other the Big Blackfoot river.[132]

Stanley's scene shows the entrance to Hellgate Canyon. Vegetation lines the river's course, and several mountains stand silently in the background. Two natives stand quietly in the foreground. Today this peaceful vista has been replaced by a modern city—Missoula, Montana. Mount Sentinel is now a landmark identifying the University of Montana, directly below.

The image's title demonstrates that explorers thought of passes in much broader terms than we do today. We would not consider Missoula a primary entrance to Cadotte Pass, which lies some ninety miles east. Rather, this image depicts the entrance to the corridor, formed by the Clark Fork and Blackfoot Rivers, that leads to the pass.

Plate LXIV. *Hell Gate—Entrance to Cadotte's Pass from the West.* Artist: Stanley.

O**N THE RETURN SEGMENT OF HIS ROUND-TRIP** from Cantonment Stevens to Fort Hall, Lt. John Mullan followed the Hell Gate (now Clark Fork) River from the Deer Lodge Valley to the Hell Gate Ronde. Stevens recorded Mullan's observations:

> After the 3d there was a great change in the weather, accompanied by snow, and the streams became frozen over, though not hard enough to bear his animal's weight, and it was therefore not so cold as it had been on Jefferson Fork. His thermometer having been broken, the exact temperature could not be determined.[133]

At one point, Mullan's party needed to cross the Hell Gate River. Although the river was frozen, the ice was not thick enough to support the weight of either horse or man. This rather dramatic image reflects the struggle of man and horse to cross the icy-cold stream. One horse is shown as having broken through the ice. Other pack animals follow a broken, jagged path from one side of the river to the other. Three horsemen are shown driving the animals. One is using a whip to encourage a horse to get up after slipping on the ice. Strangely, one man sits passively on the destination shoreline. He seems to be completely indifferent to the struggle around him, or perhaps he is just resting after successfully crossing the river.

This is a well-composed image. Leafless but lovely trees magnetize the viewer's attention. On the left is a huge pile of wood that has accumulated along the riverbank. Another interesting component, in the right-hand background, is the sawtooth peaks, etched with artistic precision and enhancement.

The Clark Fork (Hell Gate) River today —Photograph by Paul McDermott

Plate LXV. *Crossing the Hellgate River, Jan. 6th 1854.* Artist: Sohon.

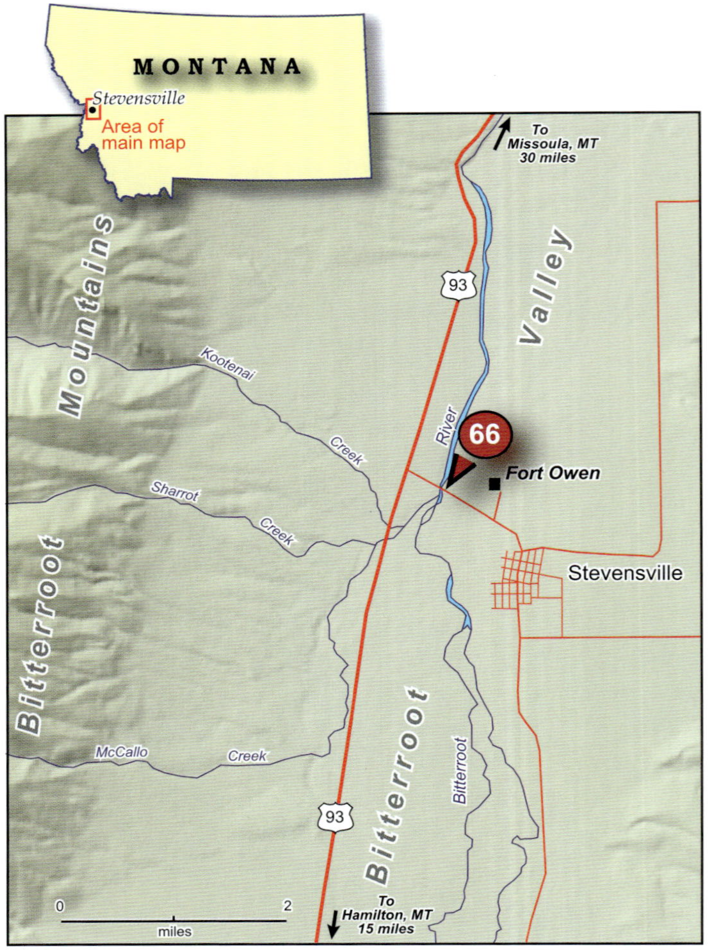

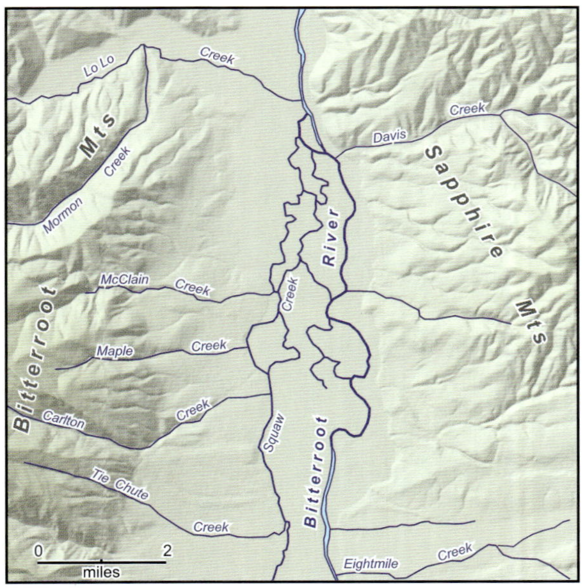

Topographic map illustrating a braided stream —Map by Philip Mobley

The Bitterroot (also known by Stevens as St. Mary's River) was originally identified by the native name *Spitlem Seukum,* meaning "waters of the bitterroot [flower]." The river begins its course near the crest of the Bitterroot Mountains in southwestern Montana and ends at the Clark Fork just west of Missoula, after flowing 120 miles. As the river descends through the valley, it changes gradient. Upon entering the valley floor between the jagged Bitterroot Mountains to the west and the lower-profile Sapphire Mountains to the east, the river creates a braided pattern, characterized by a network of intertwining channels (see map illustrating a braided stream). With braided streams, a strong central channel is sometimes difficult to determine. These streams are subject to flooding, especially during sudden snowmelts.

The river course defines a very significant component of the Montana landscape, the Bitterroot Valley, an intermontane valley characterized by relatively mild winter temperatures, with good grazing and farming land.

Stanley sketched this scene along the Bitterroot's east bank near Fort Owen. Here the fort is only a mile or so from the river. The view is toward the southeast and the Bitterroot Mountains. Several prominent mountains, referred to as the Totems, are easily identified by their sharp pointed peaks and close proximity to one another. The largest peak, today called St. Mary Peak, is in the lithograph's right center. On the bank, a local native fishes quietly.

Bitterroot River west of Fort Owen —Photograph by Paul McDermott

Plate LXVI. *Bitter Root River, near Fort Owen.* Artist: Stanley.

In Stevens's "Geographical Memoir," he provided an overview of the portion of Washington Territory after the Columbia River makes its great bend near Wallula:

> The Blue mountains, which were referred to as bordering the Walla-Walla valley on the south, have a general course westward, south of the main Columbia, until they unite with the Cascade mountains, from which flow many streams to the Columbia, the Umatilla, Willow creek, Butter creek, John Day's river, and the Deschutes river. (A sketch is here given of the junction of the Des Chutes and Columbia rivers.) On the immediate banks of the Columbia, the country is not promising; but going back a little distance the grazing is very luxuriant and excellent, and the soil rich, particularly in the river valleys. The traveler, starting from the Des Chutes river and simply passing southward a mile or two from the trail, will be struck with the freshness and luxuriance of the grass in August, September, and October, and with the arable qualities of the soil.[134]

The Deschutes is one of Oregon's great rivers. It dominates the landscape on the east side of the Cascades as it etches a path through the many basaltic lava flows in north-central Oregon. The river is 250 miles long. French trappers originally named the rapidly flowing stream *Rivière de Chutes*, meaning "river of falls."[135] Indeed, the Deschutes is a river of falls, which today translates into great white-water recreation. But during the mid-nineteenth century, pioneers attempting to cross the Deschutes had a difficult time.

According to U.S. Geological Survey data, the river has a very steady flow. The difference between the highest and lowest discharge is not great. The flow originates as meltwater that accumulates in the Crane Prairie country and collects into several lakes and reservoirs before becoming a major stream. Around Bend, Oregon, the Deschutes plummets and curls around lava outcrops emanating from volcanic cinder cones, such as Lava Butte. Later the river cuts its course into a spectacular canyon some 1,200 feet deep, characterized by nearly vertical walls. Finally, it enters the Columbia River a few miles upstream from The Dalles, Oregon.

Stanley sketched this image in November 1853. The scene provides little detail aside from defining lava bluffs on the Columbia's flanks. Water in the foreground plummets through rectangular rock blocks, moving toward the viewer. The lack of specific content suggests that Stanley made the image in a short time window. Stevens, who accompanied the artist in November 1853, was in a rush to reach Fort Vancouver.

Plate LXVII. *Columbia River—Junction of Des Chutes.* Artist: Stanley.

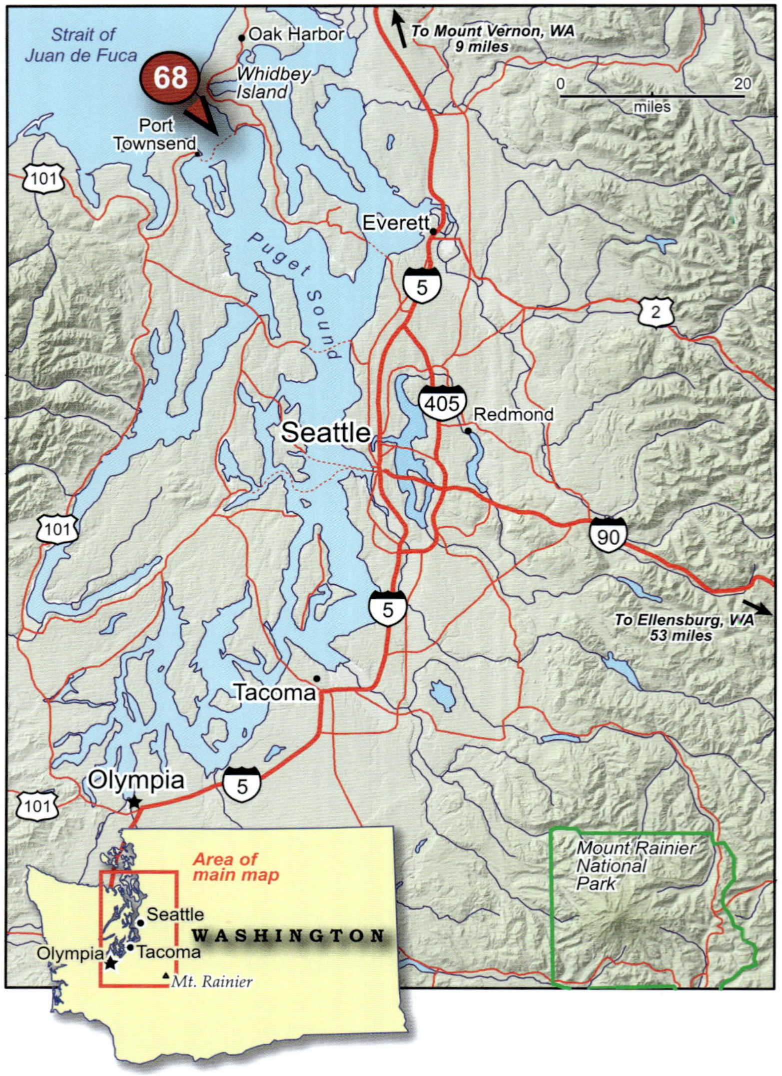

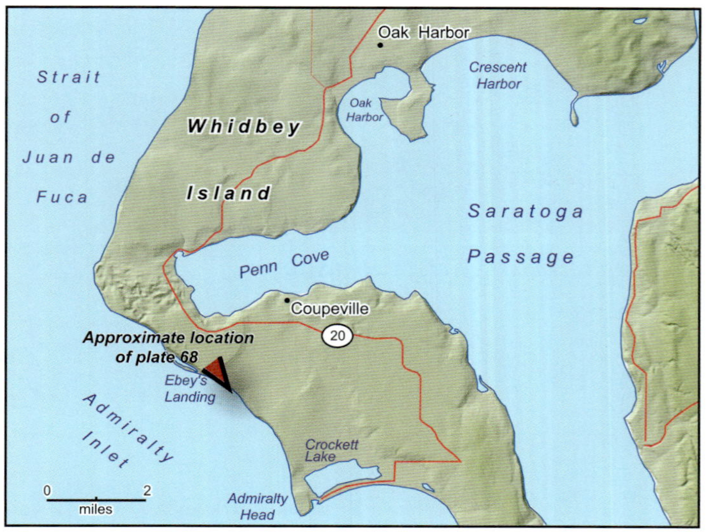

One of the final chapters of the report was devoted to a discussion of the hydrography of the coasts and the navigable waters of Washington Territory. This part of the report was prepared by Johann Georg Kohl, a German historical geographer who worked for the U.S. Coast Survey during the early 1850s. In his contribution, Kohl described various prominent hydrographic features, including Puget Sound:

> Puget Sound (see sketch) forms a most variegated compound of narrow inlets and sounds, interlinked among each other by passages and channels, and connected with Admiralty Inlet only by the above named Narrows. A number of peninsulas and islands are cut out by those different water branches.[136]

The accompanying image is intriguing for several reasons. First, one notices that the artist was captivated by rough water. Considering the fact that the image was rendered at Whidbey Island, it is likely that the depiction represents real tidal and wind activity affecting Whidbey Island's western shore, particularly near Ebey's Landing.

A turn of 90 degrees westward would yield a view of the entire Strait of Juan de Fuca. With the wind coming from the west-northwest, some very ugly wave conditions can be generated in the strait and along the island's western shore. Ebey's Landing, however, provides relatively easy access to Puget Sound; here, the slopes to the shore are low compared to the high wave-cut cliffs to the north and south.

In this image, the artist adds a little drama by depicting a canoe struggling against waves and violent underwater currents. The boats, both on- and offshore, have an elongated shape that suggests dugout logs fashioned by local natives. Mount Rainier appears in the background, but in this depiction its outline is smooth and indistinct. From this location, the mountain is about 150 miles away—its size from this distance is greatly exaggerated. Furthermore, its shape appears to be more conical than it is in reality (see also plate 69).

The lithographic copy of this view was attributed to Stanley; however, a marginal note on the manuscript watercolor indicates that Stanley prepared it from a sketch by Dr. J. G. Cooper. The original drawing was probably made during the summer of 1853, when Cooper was working as a surgeon and naturalist under Captain McClellan. Later in November 1853,

Stevens also surveyed Puget Sound in a sailboat named the *Sarah Stone*. Stevens's goals were "to visit and take a census of Indian tribes, learn something of the general character of the Sound and its harbors, and to visit Vancouver Island and its principal port, Victoria."[137]

It is truly amazing how much territory was explored or surveyed by either individuals or small groups in comparison to that covered by the main survey party. The approximate distance between Fort Snelling and Puget Sound is 3,000 miles, but our calculations determine that the entire party explored about 11,300 miles.[138] Consequently, 73 percent of the exploration was done by individuals or small groups. Remarkably, most of this exploration was completed in a year's time.

Ebey's Landing, looking southeast (toward Mount Rainier, which is not visible) —Photograph by Dave Johanson and used by permission

Plate LXVIII. *Puget Sound & Mt. Rainier from Whitby's Island.* Artist: Stanley from sketch by Dr. J. G. Cooper.

Manuscript watercolor of Puget Sound and Mount Rainier by John Mix Stanley —Courtesy of Yale University Art Gallery and used by permission

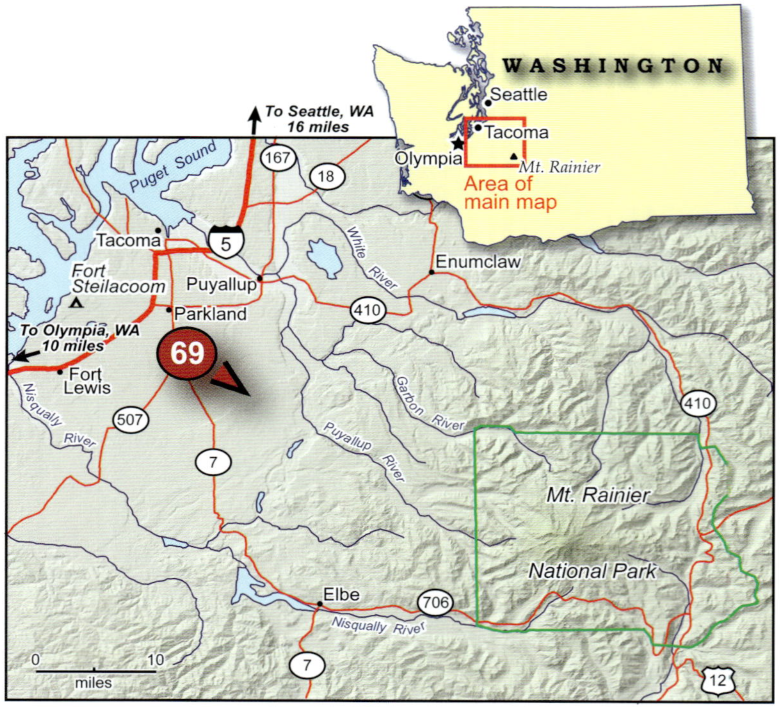

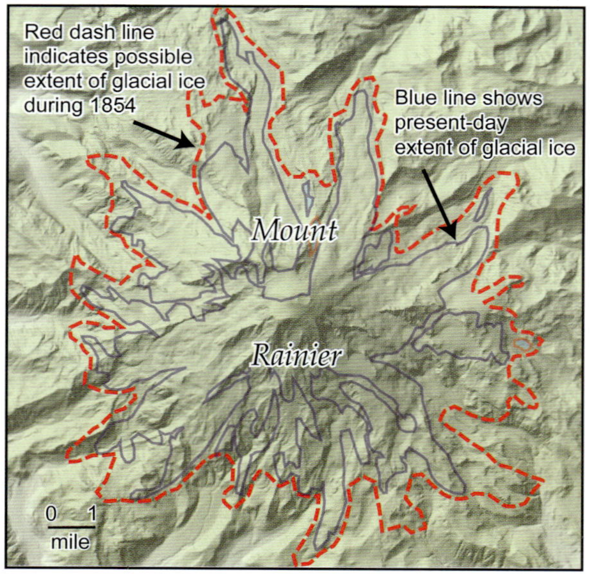

Snow and ice coverage on Mount Rainier in 1853

NEAR MOUNT RAINIER IS A MOUNTAIN PASS—Naches Pass—through which the proposed Pacific Railroad could cross. Since the peak was the most conspicuous landmark near the pass, it is not surprising that a description of it was included in Stevens's congressional report:

> Mount Rainier (see sketch) is one of the highest and most prominent peaks of the Cascade Range. It . . . seems to form the southern pillar and landmark of a large section of the Cascade range, which begins in the north with Mount Baker and branches out from this knot in several branches. . . . The mountain was first discovered by Vancouver in the beginning of May of 1792, from Port Townsend. He named it in honor of his friend, Rear Admiral Rainier.[139]

Surveyors projected a possible rail route along the Yakima and Naches Rivers to the White River or perhaps along the Yakima to the Cowlitz River, one of which the rail would follow until it reached the settlements in the Puget Sound lowlands. When railroads were finally built in the region, including the Northern Pacific in 1883 and later, small rail systems for the logging industry, each used portions of the survey's route, but a major railroad was never built.

On clear days, Mount Rainier could easily be seen from places like Fort Steilacoom, and today from Seattle and Tacoma. This mountain certainly captivated early explorers and artists with its height of 14,410 feet and the numerous valley glaciers that cover it. Today Mount Rainier is the site of a beautiful national park. There are few people in the Puget Sound region who do not stand in awe of this peak, and of the other volcanoes of the High Cascades that project their majestic cones above the lower mountains and the surrounding lowlands.

The content of Stanley's image contains two especially interesting aspects. First, the mountain is covered by more snow than is seen today. Indeed, it is now recognized that the mountain had heavier snowfall than it does today, due to snow accumulated during the cool period known as the Little Ice Age of approximately 1300 to 1850. Subsequently, the glaciers covering Mount Rainier were about 25 percent larger in area than they are today.[140]

After the 1850s, temperatures gradually began to rise, in part because of carbon dioxide produced during the Industrial Revolution. As the mountain's ice covering began melting away, the underlying structure of this stratocomposite

Plate LXIX. *Mount Rainier Viewed from near Steilacoom.* Artist: Stanley from sketch by Dr. J. G. Cooper.

volcano was exposed. Today, the layers of alternating hard and soft igneous material can easily be seen.

The second notable feature in this image is that the mountain's shape is unrealistically conical. Stanley, to whom the image is credited, probably did not see the mountain from this perspective. He used a drawing provided by another man on the expedition, Dr. J. G. Cooper. Perhaps Stanley misremembered the shape of the mountain and disbelieved the details in the sketch provided to him by Cooper. It is possible that Stanley made modifications as he prepared the manuscript watercolor and lithographic print. Alternatively, all or part of the distortion could have been in Cooper's sketch, since Mount Rainier also appears to be distorted in the previous view (see plate 68), which was also based on a sketch by Cooper.

Mount Rainier is still active and poses some danger to people living along river valleys and the adjacent Puget Sound lowland. If large amounts of snow and ice suddenly melted, lahars (mudslides) would descend rapidly through the valleys, causing major destruction in their paths.

Closer view of Mount Rainier today, showing exposed stratocomposite structure —Photograph by Paul McDermott

Mount Rainier from the west —Photograph by Paul McDermott

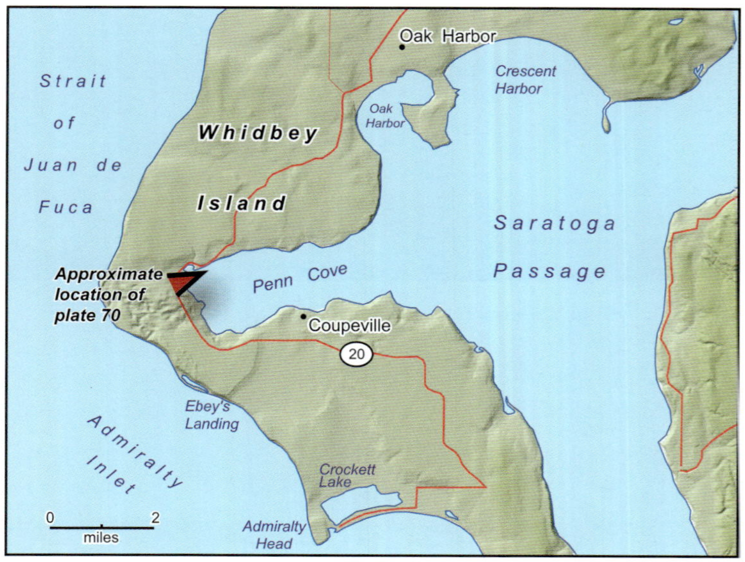

MOUNT BAKER IS ANOTHER RUGGED CASCADE VOLCANO. Its peak is 10,778 feet above sea level. Like most of the other volcanoes in the region, it is still active and capable of eruption. The last significant eruption occurred in 1870, and it was in fact actively erupting when this sketch was originally drawn in 1853.[141] Stevens did not mention the eruption in his report, however, probably because he did not want the public to know that volcanoes were still active in the Pacific Northwest, as this knowledge might discourage settlement in the region.

Mount Baker sits north of Seattle and east of Bellingham, Washington. In this particular scene, the mountain is observed in the background. The foreground features Puget Sound around the northern part of Whidbey Island, viewed from a point looking eastward from Penn Cove. Wave-cut erosion can be seen along the cove's northern bluffs.

Madrona trees —Courtesy of Dave Johanson and used by permission

Also of interest is the artist's portrayal of vegetation as part of the composition. The foreground in this print focuses on several plants, the major one being part of a Madrona tree (*Arbutus menziesii*). This tree is interesting because of its peeling red bark. Typically, it grows on steep cliffs and places were the soil is acidic and moist. Its deep roots help stabilize soil on cliff sides, minimizing erosion. Today the species is suffering from several deadly fungi. The trees have become more susceptible to infection, primarily due to a lack of rain in recent years. Furthermore, the suppression of fires has prevented young trees from sprouting.[142]

When the image was created in 1853, Mount Baker was shown erupting, but on the manuscript watercolor, a faint "x" has been placed on the volcanic plume rising from the mountain's summit, probably added by Stevens On the final lithograph the plume is missing.

Stevens's survey of Puget Sound after he arrived at Olympia was not part of the railroad survey, although he did include information about it in his official report. Stevens's comments on Mount Baker conclude the survey report:

> [It] is one of the loftiest and most conspicuous peaks of the northern Cascade range; it is nearly as high as Mount Rainier and, like that mountain, its snow-covered pyramid has the form of a sugar-loaf. (see sketch.) It is visible from all the waters and islands which we have just described, and from the whole southeastern part of the Gulf of Georgia, and likewise from the eastern division of the Strait of De Fuca. It is for this region a natural and important landmark, as Mount Rainier is for Admiralty Inlet. . . . Mt. Baker, which stands quite near the 49th degree of north latitude, may be said to be the northwestern boundary pillar of the territory of the United States, and with it we conclude this review.[143]

Mount Baker and the Cascade Mountains from Orcas Island —Photograph by Paul McDermott

Plate LXX. *Mt. Baker & Cascade Range, from Whitby's Island.* Artist: Stanley from sketch by Dr. J. G. Cooper.

Manuscript watercolor of Mount Baker by John Mix Stanley —Courtesy of Yale University Art Gallery and used by permission

Notes

[1] William Goetzmann, in his book *Exploration and Empire: The Explorer and the Scientist in Winning the American West* (New York: Alfred A. Knopf, 1966), 281, suggested that the placement of the surveys in the hands of the Secretary of War was "a stratagem to take the decision out of the hands of an indecisive Congress."

[2] Goetzmann, *Exploration and Empire*, 281, noted that this selection was not unbiased. "It was in Stevens' direct interest to report favorably on the northern route," since he was concurrently the governor of Washington Territory.

[3] Robert Taft, *Artists and Illustrators of the Old West, 1850–1900* (New York: Charles Scribner, 1953), 5.

[4] For a description and discussion of the maps associated with the Pacific Railroad Surveys, see Carl I. Wheat, *Mapping the Trans-Mississippi West*, vol. 4 of *From the Pacific Railroad Surveys to the Onset of the Civil War, 1855–1860* (San Francisco: Institute of Historical Cartography, 1960), 1–28, 67–91.

[5] Taft, *Artists and Illustrators,* 276.

[6] http://www.fada.com/browse_by_artist.html?gallery_no=26 & artist=3921&bio=1.

[7] John Mullan, *Report on the Construction of a Military Road from Fort Walla-Walla to Fort Benton* (Washington: U.S. Government Printing Office, 1863), published separately as Senate Executive Document 43, 37th Cong., 3rd Sess., Serial 1149; and Hazard Stevens, *The Life of Isaac Ingalls Stevens* (Boston: Houghton Mifflin, 1900).

[8] Paul D. McDermott and Ronald E. Grim, comps., *Gustavus Sohon's Cartographic and Artistic Works: An Annotated Bibliography* (Washington, DC.: Library of Congress, Philip Lee Phillips Society, 2002).

[9] William H. Goetzmann, *Looking at the Land of Promise: Pioneer Images of the Pacific Northwest* (Pullman: Washington State University Press, 1988), 33. Several of Sohon's drawings appear on pages 71–74.

[10] The twelve volumes of reports are collectively known by the title *Reports of Explorations and Surveys, to Ascertain the Most Practicable and Economic Route for a Railroad from the Mississippi River to the Pacific Ocean Made under the Direction of the Secretary of War in 1853–54*. Stevens's preliminary report, published as part of volume 1, is entitled "Report of Explorations for a Route for the Pacific Railroad, near the Forty-Seventh and Forty-Ninth Parallels of North Latitude from St. Paul to Puget Sound." His final report, published both as a supplement to volume 1 and as a two-part volume 12, was entitled "Narrative and Final Report of Explorations for a Route for a Pacific Railroad near the Forty-Seventh and Forty-Ninth Parallels of North Latitude from St. Paul to Puget Sound." Since most volumes of these reports were published both as House and Senate documents and at various times, producing a complex and lengthy bibliographic description, the interested reader is referred to Taft, 254–55, for a full bibliographic listing.

[11] Although only one route map was published with Stevens's final report, two sets of three maps each were published with earlier versions of the report; see Wheat, *From the Pacific Railroad Surveys to the Onset of the Civil War,* 208–10.

[12] The version of G. K. Warren's map is from the collections of the Geography and Map Division of the Library of Congress; it is reproduced digitally at http://memory.loc.gov/ammem/gmdhtml/rrhtml/rrintro.html.

[13] Isaac I. Stevens, *Reports of Explorations and Surveys to Ascertain the Most Practicable and Economic Route for a Railroad from the Mississippi River to the Pacific Ocean,* 36th Cong., 1st Sess., H. Exec. Doc. 56 (Washington, DC: GPO, 1860), 35.

[14] Ibid., 36.

[15] Ibid., 39.

[16] http://mnplaces.mnhs.org/upham/index.cfm. This Web site is very useful for information about place-name origins in Minnesota. It is based on Warren Upham's *Minnesota Place Names,* originally published in 1920.

[17] I. Stevens, *Reports of Explorations and Surveys* (1860), 45. The party also used India rubber boats to cross the Chippewa River.

[18] http://mnplaces.mnhs.org/upham/index.cfm.

[19] I. Stevens, *Reports of Explorations and Surveys* (1860), 46.

[20] http://mnplaces.mnhs.org/upham/index.cfm.

[21] I. Stevens, *Reports of Explorations and Surveys* (1860), 48.

[22] Ibid., 314.

[23] http://mnplaces.mnhs.org/upham/index.cfm.

[24] I. Stevens, *Reports of Explorations and Surveys* (1860), 49.

[25] Ibid., 50.

[26] Ibid., 53.

[27] Ibid., 57.

[28] H. Stevens, *Life of Isaac Ingalls Stevens,* 326.

[29] http://esvc000736.wic009u.server-web.com/background/suspects.htm.

[30] I. Stevens, *Reports of Explorations and Surveys* (1860), 58.

[31] Anthony R. Orme, *The Physical Geography of North America* (Oxford: Oxford University Press, 2000), 41.

[32] Ibid., 59.

[33] http://home.att.net~mman/FortUnion.htm.

[34] I. Stevens, *Reports of Explorations and Surveys* (1860), 60.

[35] Ibid., 63–64.

[36] Ibid., 65.

[37] Ibid., 65–67.

[38] Ibid., 66.

[39] https://www.mylocalgov.com/wellscountynd/index.asp.

[40] I. Stevens, *Reports of Explorations and Surveys* (1860), 68.

[41] Ibid., 73.

[42] Edwin Thompson Denig, *The Assiniboine* (Norman: University of Oklahoma Press, 2000).

[43] I. Stevens, *Reports of Explorations and Surveys* (1860), 74.

[44] Francis J. Monkhouse, *A Dictionary of Geography* (Chicago: Aldine Publishing, 1965), 81.

[45] I. Stevens, *Reports of Explorations and Surveys* (1860), 84.

[46] Ian Frazier, *Great Plains* (New York: Farrar Straus Giroux, 1989), 19.

[47] For example, an average year could involve 150,000 buffalo skins traded for goods. Erin Thompson, *Fort Union Trading Post: Fur Trade Empire on the Upper Missouri* (Williston, ND: Fort Union Association, 2003); *Fort Union Trading Post: Historical Structures Report, Part 2* (Washington, DC: U.S. National Park Service, Historical Data Section Division of History Office of Archeology and Historic Preservation, 1968).

[48] I. Stevens, *Reports of Explorations and Surveys* (1860), 85–86.

[49] Joel Overholser, *Fort Benton: World's Innermost Port* (Fort Benton, MT: Joel Overholser, 1987), 21.

[50] I. Stevens, *Reports of Explorations and Surveys* (1860), 89. There is no evidence that the survey party used the wagons and carts beyond Fort Benton.

[51] Ibid., 90.

[52] Gary E. Moulton, ed. *The Journals of the Lewis and Clark Expedition* (Lincoln: University of Nebraska Press, 1987), 4:124.

[53] The Milk River stretches from its origin in today's Glacier National Park and moves generally eastward for some 625 miles. For about 100 miles, or 20 percent of its course, the river flows through Canada, emerging into the United States to continue its descent to the Missouri, which it joins near Fort Peck, Montana.

[54] I. Stevens, *Reports of Explorations and Surveys* (1860), 90.

[55] http://www.coopext.colostate.edu/4dmg/Trees/cotton.htm.

[56] I. Stevens, *Reports of Explorations and Surveys* (1860), 91.

[57] Ibid., 94.

[58] Ibid., 95.

[59] http://cometobservation.com/bright_comet.html#pre1900 provides information about comets from 1800 to 1899.

[60] This picture was created using Sky6 software, produced by Bisque, with the additional assistance of Dr. William Davis, National Institute of Science and Technology, Gaithersburg, Maryland.

[61] I. Stevens, *Reports of Explorations and Surveys* (1860), 97.

[62] Ibid.

[63] Ibid.

[64] This mountain complex is composed of igneous intrusions that resulted from the injection of molten rock into the earth's crust. See David Alt and Donald W. Hyndman, *Roadside Geology of Montana* (Missoula, MT: Mountain Press, 1986).

[65] I. Stevens, *Reports of Explorations and Surveys* (1860), 101.

[66] H. Stevens, *Life of Isaac Ingalls Stevens,* 362.

[67] Ibid., 373.

[68] Overholser, *Fort Benton*.

[69] I. Stevens, *Reports of Explorations and Surveys* (1860), 103–4.

[70] Ibid., 107.

[71] Ibid., 106.

[72] Ibid., 98.

[73] Ibid., 107.

[74] Rick and Susie Graetz, *This Is Montana* (Helena: Northern Rockies Publishing, 2003).

[75] I. Stevens, *Reports of Explorations and Surveys* (1860), 111. On September 6, at the direction of Stevens, survey party engineer A. W. Tinkham ascended all three buttes and took barometric measurements to determine their heights. He found all three to be about 6,700 feet above sea level.

[76] The Siksika wore black moccasins. The Kainah, or Blood, peoples used red clay as a body decoration. The Piegans owned more

horses than did the other Blackfeet groups. See John C. Ewers, *The Blackfeet: Raiders on the Northwestern Plains* (Norman: University of Oklahoma Press, 1958).

[77] I. Stevens, *Reports of Explorations and Surveys* (1860), 117.

[78] H. Stevens, *Life of Isaac Ingalls Stevens*, 376–77.

[79] Ibid., 382.

[80] In 1853 portions of the Clark Fork had entirely different names—Hell Gate upstream from today's Missoula, Montana, and Bitter Root downstream toward the northwest. For a time, the Bitterroot, south of Missoula, was called St. Mary's, or St. Marie's, River.

[81] I. Stevens, *Reports of Explorations and Surveys* (1860), 125.

[82] H. Stevens, *Life of Isaac Ingalls Stevens,* 384.

[83] Ibid., 385.

[84] I. Stevens, *Reports of Explorations and Surveys* (1860), 131.

[85] Ibid., 133.

[86] H. Stevens, *Life of Isaac Ingalls Stevens,* 389.

[87] Sohon visited the mission in 1858 and 1859 and made additional drawings of it. This one is in pencil and the other, now owned by Georgetown University, is in color. See Paul D. McDermott and Ronald E. Grim, *Gustavus Sohon's Cartographic and Artistic Works: An Annotated Bibliography,* Philip Lee Phillips Society Occasional Paper Series, no. 4 (Washington, DC: Library of Congress, Geography and Map Division, 2002), 42–43, 49; and McDermott and Grim, "Images of Colonel George Wright's Campaign of 1858: The Artistic Views of Gustavus Sohon," *Columbia, the Magazine of Northwest History* 16, no. 2 (Summer 2002), 16–22.

[88] I. Stevens, *Reports of Explorations and Surveys* (1860), 135–36.

[89] An original drawing of the falls by Gustavus Sohon is now owned by Washington State University in Pullman. Sohon drew the image when he accompanied Col. George Wright's military expedition in 1858. See McDermott and Grim, *Gustavus Sohon's Cartographic and Artistic Works,* 42; and McDermott and Grim, "Images of Colonel George Wright's Campaign of 1858," 22.

[90] There were two Fort Colvilles in the nineteenth century. The Hudson's Bay Company post was located on the Columbia River. The other one was created as a military post in 1859; it was about three miles northeast of present-day Colville, Washington.

[91] I. Stevens, *Reports of Explorations and Surveys* (1860), 147.

[92] H. Stevens, *Life of Isaac Ingalls Stevens,* 399.

[93] I. Stevens, *Reports of Explorations and Surveys* (1860), 151.

[94] Unfortunately, the site was never fully investigated, and the shelter now lies beneath forty feet of water behind the Lower Monumental Dam. Most of our knowledge about the occupancy there was gained from a study in 1965. See http://www.kennewick-man.com/kman/series/story/3275027p-3304646c.html. See also Ruth Kirk and Richard D. Daugherty, *Archaeology in Washington* (Seattle: University of Washington Press, 2007), 18–26.

[95] I. Stevens, *Reports of Explorations and Surveys* (1860), 151.

[96] McDermott and Grim, *Gustavus Sohon's Cartographic and Artistic Works,* 47; McDermott and Grim, "Early Images of the Northwest: The Works of Gustav Sohon," *Columbia, the Magazine of Northwest History* (Winter 2007–8), 20–21. Sometime in 1858 or 1859, Sohon made an oil painting of the falls, now in the possession of his descendants.

[97] I. Stevens, *Reports of Explorations and Surveys* (1860), 151.

[98] Writers Program (Washington), *Washington: A Guide to the Evergreen State* (Portland, OR: Binfords and Mort, 1941; rev. ed., 1950), 369. The first fort built at this location was Fort Nez Perce (1818). After the Hudson's Bay Company acquired the property (1821), the post was rebuilt and renamed Fort Walla Walla. Fire destroyed the wooden structure in 1841, and it was rebuilt again using adobe. This did not prevent it from burning down again in 1855 during a conflict between settlers and the natives. Stanley depicted the adobe building, which he visited first in 1847 and again in 1853.

[99] I. Stevens, *Reports of Explorations and Surveys* (1860), 155.

[100] http://www.ecotrust.org/community/recalling celilo.html.

[101] I. Stevens, *Reports of Explorations and Surveys* (1860).

[102] http://www.nps.gov/fova/index.htm.

[103] McDermott and Grim, *Gustavus Sohon's Cartographic and Artistic Works,* 3, 6.

[104] I. Stevens, *Reports of Explorations and Surveys* (1860), 155.

[105] http://www.historylink.org/essays/output.cfm?file_id=7797.

[106] http://www.lewis-clark.org/content/content-article.asp?ArticleID=2728. This site has interesting maps created by Lewis and Clark showing the Cascades.

[107] I. Stevens, *Reports of Explorations and Surveys* (1860).

[108] http://www.historylink.org/essays/output.cfm?file_id=5102. Another Web site, http://www.ghosttownsusa.com/kettlefalls.htm, discusses another possible Native American name.

[109] I. Stevens, *Reports of Explorations and Surveys* (1860), 160.

[110] Ibid., 169.

[111] The army attempted to defeat and capture Joseph's group of about 750 Nez Perces, who were fleeing to Canada to escape government relocation. After a surprise early-morning attack, the battle lasted about a day and a half. By the time it ended, 156 people were either killed or wounded, with both sides suffering roughly the same number of causalities. The victory enabled the tribe to move

farther toward the Canadian border, though Joseph and most of his people never made it across, the army having defeated them in the October 1877 Battle of the Bear Paw Mountains. One government report stated that 150 Nez Perces managed to slip away and cross to Canada. See Robert R. McCoy, *Chief Joseph, Yellow Wolf, and the Creation of Nez Perce History in the Pacific Northwest* (New York: Routledge, 2004): 120, 123; and www.nps.gov/biho/, the Web site for Big Hole Battlefield.

112 http://www.trails.com/tcatalog_trail.asp?trailid=HGR211-013.

113 I. Stevens, *Reports of Explorations and Surveys* (1860), 172.

114 John Small and Michael Witherick, *A Modern Dictionary of Geography* (London: Edward Arnold, 1986), 154.

115 I. Stevens, *Reports of Explorations and Surveys* (1860), 178.

116 Carlene Cross, *The Undying West: A Chronicle of Montana's Camas Prairie* (Golden, CO: Fulcrum Publishing, 1999), 11.

117 H. Stevens, *Life of Isaac Ingalls Stevens,* 386.

118 I. Stevens, *Reports of Explorations and Surveys* (1860), 178.

119 Ibid., 178.

120 Ibid., 179.

121 Ibid., 180.

122 http://visitmt.com/history/Montana_the_Magazine_of_Western History/winter1004/lolo.htm.

123 I. Stevens, *Reports of Explorations and Surveys* (1860), 180.

124 Ibid.

125 Moulton, *Journals of the Lewis and Clark Expedition,* 4:203; Noah Brooks, *First Across the Continent: The Story of the Exploring Expedition of Lewis and Clark in 1803–4–5* (New York: Charles Scribner's Sons, 1901), 309.

126 http://visitmt.com/history/Montana_the_Magazine_of_Western History/winter1004/lolo.htm.

127 I. Stevens, *Reports of Explorations and Surveys* (1860), 181.

128 Ibid., 173.

129 Ibid., 200.

130 Ibid., 214.

131 Ibid., 239.

132 Ibid., 243–44.

133 Ibid., 172.

134 Ibid., 257.

135 John Muir, *Steep Trails,* ed. William Frederic Badè (Boston: Houghton Mifflin, 1918), 339.

136 I. Stevens, *Reports of Explorations and Surveys* (1860), 289.

137 H. Stevens, *Life of Isaac Ingalls Stevens,* 409.

138 This mileage calculation does not include Stevens's survey of Puget Sound after he arrived at Olympia.

139 I. Stevens, *Reports of Explorations and Surveys* (1860), 290–91.

140 Brian Fagan, *The Little Ice Age: How Climate Made History, 1300–1850* (New York: Basic Books, 2000).

141 http://vulcan.wr.usgs.gov/Volcanoes/Baker/description_baker.html.

142 *Seattle Times,* September 30, 2004: http://seattletimes.nwsource.com/html/outdoors/2002049584_nwwwmadrona30.html.

143 I. Stevens, *Reports of Explorations and Surveys* (1860), 306.

Bibliography

Alt, David D., and Donald W. Hyndman. *Roadside Geology of Montana*. Missoula, MT: Mountain Press, 1986.

Botkin, Daniel B. *Passage of Discovery: The American Rivers Guide to the Missouri River of Lewis and Clark*. New York: Perigee Books, 1999.

Brewer, William H. *Up and Down California in 1860–1864: The Journal of William Brewer*. Edited by Francis P. Farquhar. New Haven: Yale University Press, 1930.

Brooks, Noah. *First across the Continent: The Story of the Exploring Expedition of Lewis and Clark in 1803–4–5*. New York: Charles Scribner's Sons, 1901.

Bushnell, David I. "John Mix Stanley, Artist-Explorer." In *Smithsonian Institution, Annual Report, 1924*. Washington, DC: U.S. Government Printing Office, 1925: 507–12.

Casler, Michael M. *Steamboats of the Fort Union Fur Trade: An Illustrated Listing of Steamboats on the Upper Missouri River, 1831–1867*. Williston, ND: Fort Union Association, 1999.

Coates, Robert M. *The Outlaw Years: The History of the Land Pirates of the Natchez Trace*. 1930. Reprint. Lincoln: University of Nebraska Press, 1986.

Cohen, Paul. *Mapping the West: America's Westward Movement, 1524–1890*. New York: Rizzoli, 2002.

Congressional Information Service. *CIS Serial Set Index, Part 14, Index and Carto-Bibliography of Maps, 1789–1969*. Vol. 1, *Carto-Bibliography of Maps, Serial Set Volumes ASP01-2160*. Bethesda, MD: Congressional Information Service, 1995.

Cross, Carlene. *The Undying West: A Chronicle of Montana's Camas Prairie*. Golden, CO: Fulcrum Publishing, 1999.

Denig, Edwin Thompson. *The Assiniboine*. Norman: University of Oklahoma Press, 2000.

———. *Five Indian Tribes of the Upper Missouri: Sioux, Arickaras, Assiniboines, Crees, Crows*. Norman: University of Oklahoma Press, 1972.

Drury, Clifford Merrill, ed. *On to Oregon: The Diaries of Mary Walker and Myra Eells.* Lincoln: University of Nebraska Press, 1998.

Ewers, John C. *Artists of the Old West*. Garden City, NY: Doubleday, 1965.

———. *The Blackfeet: Raiders on the Northwestern Plains*. Norman: University of Oklahoma Press, 1958.

———. *Gustavus Sohon's Portraits of Flathead and Pend d'Oreille Indians*. Smithsonian Miscellaneous Collections. Vol. 110, no.7. Washington, DC: Smithsonian Institution, 1948.

———. *The Horse in Blackfoot Indian Culture, with Comparative Material from Other Western Tribes*. Washington, DC: U.S. Government Printing Office, 1955.

———. *Indian Life on the Upper Missouri*. Norman: University of Oklahoma Press, 1968.

———. *Plains Indian History and Culture: Essays on Continuity and Change*. Norman: University of Oklahoma Press, 1997.

Fagan, Brian. *The Little Ice Age: How Climate Made History, 1300–1850*. New York: Basic Books. 2000.

Frazier, Ian. *Great Plains*. New York: Farrar Straus Giroux, 1989.

Friis, Herman R. "Original and Published Sources in Research in Historical Geography: A Comparison." In *Pattern and Process: Research in Historical Geography*. Edited by Ralph E. Ehrenberg. Washington, DC: Howard University Press, 1975: 139–59.

Foster, John E., Dick Harrison, and I. S. MacLaren, eds. *Buffalo*. Edmonton: University of Alberta Press, 1992.

Geist, Valerius. *Buffalo Nation: History and Legend of the North American Bison*. Stillwater, MN: Voyageur Press, 1996.

Goetzmann, William H. *Army Exploration in the American West, 1803–1863*. New Haven: Yale University Press, 1959.

———. *Exploration and Empire: The Explorer and the Scientist in the Winning of the American West*. New York: Alfred A. Knopf, 1966.

———. *Looking at the Land of Promise: Pioneer Images of the Pacific Northwest*. Pullman: Washington State University Press, 1988.

Graetz, Rick and Susie. *This Is Montana*. Helena, MT: Northern Rockies Publishing, 2003.

Haines, Francis. *The Buffalo: The Story of American Bison and Their Hunters from Prehistoric Times to the Present*. Norman: University of Oklahoma Press, 1995.

Hazard, Joseph T. *Companion of Adventure: A Biography of Isaac Ingalls Stevens*. Portland, OR: Binfords & Mort, 1952.

Hedren, Paul L., ed. *Forty Years a Fur Trader on the Upper Missouri: The Personal Narrative of Charles Larpenteur, 1833–1872*. Lincoln: University of Nebraska Press, 1989.

Holmes, Kenneth L., ed. *Covered Wagon Women: Diaries and Letters from the Western Trails, 1852*. 11 vols. Lincoln: University of Nebraska Press, 1983–2000.

Hunter, Louis C. *Steamboats on the Western Rivers: An Economic and Technological History*. New York: Dover Publishing, 1993.

Kinietz, W. Vernon. *John Mix Stanley and His Indian Paintings*. Ann Arbor: University of Michigan Press, 1942.

Kirk, Ruth, and Richard D. Daugherty. *Archaeology in Washington*. Seattle: University of Washington Press, 2007.

MacLean, Norman. *A River Runs Through It*. New York: Pocket Books, 1976.

Mayhew, Susan, and Anne Penny. *The Concise Oxford Dictionary of Geography*. New York: Oxford University Press, 1992.

McCoy, Robert R. *Chief Joseph, Yellow Wolf, and the Creation of Nez Perce History in the Pacific Northwest*. New York: Routledge, 2004.

McDermott, Paul D., and Ronald E. Grim. "Early Images of the Northwest: The Works of Gustav Sohon." In *Columbia, the Magazine of Northwest History* 21, no. 4 (Winter 2007–8): 16–22.

———. *Gustavus Sohon's Cartographic and Artistic Works: An Annotated Bibliography*. Philip Lee Phillips Society Occasional Paper Series, no. 4. Washington, DC: Library of Congress, Geography and Map Division, 2002.

———. "Images of Colonel George Wright's Campaign of 1858: The Artistic Views of Gustavus Sohon." In *Columbia, the Magazine of Northwest History* 16, no. 2 (Summer 2002): 16–22.

———. "The Mapmaker as Artist: Early Visuals of the American West." In *Mercator's World* 1, no. 1 (1996): 28–33.

———. "Maps of the Mullan Road." In *Proceedings of the American Congress on Surveying and Mapping, 36th Annual Meeting, February 22–28, 1976*. Washington, DC: American Congress on Surveying and Mapping, 1976: 212–26.

Meinig, Donald W. *The Great Columbia Plain: A Historical Geography, 1805–1910*. Seattle: University of Washington Press, 1968.

———. "Isaac Stevens: Practical Geographer of the Early Northwest." In *Geographical Review* 45 (October 1955): 542–58.

Middleton, W. E. Knowles. *A History of the Barometer*. Baltimore: Johns Hopkins Press, 1964.

———. *A History of the Thermometer and Its Use in Meteorology*. Baltimore: Johns Hopkins Press, 1966.

Monkhouse, Francis J. *A Dictionary of Geography*. Chicago: Aldine Publishing, 1965.

Moulton, Gary E., ed. *The Journals of the Lewis and Clark Expedition*. 13 vols. Lincoln: University of Nebraska Press, 1983–2001.

Muir, John. *Steep Trails*. Edited by William Frederic Badè. Boston: Houghton Mifflin, 1918.

Mullan, John. *Report on the Construction of a Military Road from Fort Walla-Walla to Fort Benton.* Washington, DC: U.S. Government Printing Office, 1863. Published separately as Senate Executive Document 43, 37th Congress, 3rd Session, Serial 1149.

National Geographic Society, Special Publications Division. *Great American Journeys*. Washington, DC: National Geographic Society, 1989.

Nicandri, David. *Northwest Chiefs: Gustave Sohon's Views of the 1855 Stevens Treaty Councils*. Tacoma: Washington State Historical Society, 1986.

———. "Isaac Stevens and the Expeditionary Artists of the Northern West." In *Encounters with a Distant Land: Exploration and the Great Northwest*. Edited by Carlos A. Schwantes and Evelyne Stitt Pickett. Moscow: University of Idaho Press, 1994: 133–48.

Orme, Antony R. *The Physical Geography of North America.* Oxford: Oxford University Press, 2000.

Overholser, Joel. *Fort Benton: World's Innermost Port*. Fort Benton, MT: Joel Overholser, 1987.

Rau, Weldon W. "The Columbia River Gorge and Early Emigrant Travel: Through the Magnificent Gateway." In *Columbia, the Magazine of Northwest History* 15, no. 4 (Winter 2001–2): 24–29.

Reinhartz, Dennis, and Gerald D. Saxon, eds. *Mapping and Empire: Soldier-Engineers on the Southwestern Frontier*. Austin: University of Texas Press, 2005.

Richards, Kent D. *Isaac I. Stevens: A Young Man in a Hurry.* Provo, UT: Brigham Young University Press, 1979.

———. "The Young Napoleons: Isaac I. Stevens, George B. McClellan and the Cascade Route." In *Columbia, the Magazine of Northwest History* 3, no. 4 (Winter 1989–90): 21–23.

Schoolcraft, Henry Rowe. *Algic Researches*. New York: Harper & Brothers, 1839.

———. *Historical and Statistical Information Respecting the History, Condition, and Prospects of the Indian Tribes of the United States*. 6 vols. Philadelphia: Lippincott, Grambo, 1851–57.

Small, John, and Michael Witherick. *A Modern Dictionary of Geography*. London: Edward Arnold, 1986.

Stevens, Hazard. *The Life of Isaac Ingalls Stevens.* 2 vols. Boston: Houghton Mifflin, 1900.

Stevens, Isaac I. *Narrative and Final Report of Explorations for a Route for a Pacific Railroad, near Forty-seventh and Forty-ninth Parallels of North Latitude from St. Paul to Puget Sound,* Vol. 12, bk. 1 of *Report of Explorations and Surveys to Ascertain the Most Practicable and Economic Route for a Railroad from the Mississippi River to the Pacific Ocean*. 36th Cong., 1st Sess., House Executive Document 56. Washington, DC: U.S. Government Printing Office, 1855.

———. *Botanical Report* [and] *Zoological Report,* Vol. 12, bk. 2 of *Reports of Explorations and Surveys to Ascertain the Most Practicable and Economic Route for a Railroad from the Mississippi River to the Pacific Ocean*. 36th Cong., 1st Sess., House Executive Document 56. Washington, DC: U.S. Government Printing Office, 1860.

Taft, Robert. *Artists and Illustrators of the Old West, 1850–1900*. New York: Charles Scribner, 1953.

Thompson, Erin N. *Fort Union Trading Post: Fur Trade Empire on the Upper Missouri*. Williston, ND: Fort Union Association, 2003.

———. *Fort Union Trading Post: Historical Structures Report, Part 2.* Washington DC: U.S. National Park Service, Historical Data Section Division of History Office of Archeology and Historic Preservation, 1968.

Upham, Warren. *Minnesota Place Names: A Geographical Encyclopedia*. 3rd ed. St. Paul: Minnesota Historical Society Press, 2001.

Wheat, Carl I. *From the Pacific Railroad Surveys to the Onset of the Civil War, 1855–1860*. Vol. 4 of *Mapping the Transmississippi West*. San Francisco: Institute of Historical Cartography, 1960.

Williamson, Robert Stockton. *On the Use of the Barometer on Surveys and Reconnaissances*. New York: D. Van Nostrand, 1868.

Wishart, David J. *The Fur Trade of the American West, 1807–1840: A Geographical Synthesis*. Lincoln: University of Nebraska Press, 1992.

Wolter, John A. "Johann Georg Kohl in America." In *Progress of Discovery: Johann Georg Kohl*. Edited by Hans-Albrecht Koch, Margrit B. Krewson, and John A. Wolter. Graz, Austria: Akademische Druck-und Verlagsanstalt, 1993: 133–59.

Writers Program (Washington). *Washington: A Guide to the Evergreen State*. Rev. ed. Portland, OR: Binfords & Mort, 1950.

Illustrations Index

Note: Page numbers for images by Stanley are in red, by Sohon are in blue. Page numbers for text discussions of images follow those for images.

General Index

About the Authors

Paul D. McDermott completed graduate studies in physical geography, cartography, and historical geography at the Universities of Washington and Maryland. He taught geography and cartography for thirty-four years at Montgomery College in Rockville, Maryland. Having grown up in Washington, he has had a lifelong interest in the historical geography of the Pacific Northwest, particularly Isaac Stevens's northern railroad survey and the Mullan Military Road. He is also an avid photographer.

A resident of Maryland, McDermott has written and lectured extensively on the historical geography of western Maryland and the Chesapeake & Ohio Canal. Along with Ronald Grim, he has written numerous papers, including articles published by *Mercator's World,* the Washington State Historical Society, and the Library of Congress Philip Lee Phillips Society.

Ronald E. Grim was trained in cultural and historical geography at the University of Maryland. He worked for ten years in the Cartographic and Architectural Division at the National Archives and Records Service, and twenty-two years in the Geography and Map Division at the Library of Congress. In the latter position, he curated five major exhibits, including *Rivers, Edens, Empires: Lewis & Clark and the Revealing of America*. Currently the curator of maps at the Norman B. Leventhal Map Center at the Boston Public Library, he lives in Boston.

Grim has written extensively on historical geography and the history of American cartography during the eighteenth and nineteenth centuries. He currently serves as the book review editor for *Imago Mundi,* the international journal of the history of cartography.

Philip Mobley, a professional cartographer, was educated at Kutztown and Akron Universities, with concentrations in geography and cartography. An adjunct professor of geography at George Mason University, he has also taught cartography and geography courses at Montgomery College and George Washington University. He was the co-cartographer for maps published in Warren Hofstra's *The Planting of New Virginia* (2004) and McDermott and Grim's *Gustavus Sohon's Cartographic and Artistic Works* (2002). He resides in Virginia.